OPTIMUM DESIGN OF METAL STRUCTURES

OPTIMUM DESIGN OF METAL STRUCTURES

JÓZSEF FARKAS, D.Sc.
Professor of Metal Structures
Technical University of Miskolc
Hungary

Translation Editor:
MILIJA N. PAVLOVIĆ, M.Sc., Ph.D., B.Eng.
Department of Civil Engineering
Imperial College of Science and Technology
University of London

ELLIS HORWOOD LIMITED
Publishers · Chichester

Halsted Press: a division of
JOHN WILEY & SONS
New York · Brisbane · Chichester · Ontario

First published in English in 1984 by

ELLIS HORWOOD LIMITED, PUBLISHERS
Market Cross House, Cooper Street, Chichester West Sussex, England

and

AKADÉMIAI KIADÓ, Budapest

Distributors:

Australia, New Zealand, South-east Asia:
Jacaranda-Wiley Ltd., Jacaranda Press,
JOHN WILEY & SONS INC.,
G.P.O. Box 859, Brisbane, Queensland 4001, Australia

Canada:
JOHN WILEY & SONS CANADA LIMITED
22 Worcester Road, Rexdale, Ontario, Canada.

Europe, Africa:
JOHN WILEY & SONS LIMITED
Baffins Lane, Chichester, West Sussex, England.

*East European Countries, China, People's Republic of Vietnam,
Korean People's Republic, Cuba and Mongolia:*
AKADÉMIAI KIADÓ, Budapest

North and South America and the rest of the world:
Halsted Press: a division of
JOHN WILEY & SONS
605 Third Avenue, New York, N.Y. 10016, U.S.A.

© 1984 Akadémiai Kiadó, Budapest

British Library Cataloguing in Publication Data
Farkas, J.
Optimum design of metal structures.
1. Structural design
I. Title
624.1'82   TA695

Library of Congress Card No. 83-10870

ISBN 0-85312-497-3 (Ellis Horwood Ltd., Publishers)
ISBN 0-470-27482-4 (Halsted Press)

Printed in Hungary

Table of Contents

Author's Preface

In recent years the development of structural design has been considerable. Based on the ready availability of sophisticated techniques of modern structural analysis, structural synthesis advanced rapidly. This synthesis elaborates the analytical results at a higher level and its aim is to find an optimum solution.

An optimum may be defined in various ways. According to the most widely accepted definition *the optimum structure satisfies the design constraints and its mass or cost is minimal.* Thus, in the optimum design procedure, the cost function and the design constraints are to be defined on the basis of analytical results and an optimum solution should be searched which minimizes the cost function and satisfies the design constraints.

The minimum weight (mass) design was developed in aircraft construction. *The minimum cost design,* however, gives more realistic solutions in other fields, e.g. in the design of mechanical or civil engineering systems.

First, the optimum dimensions of the simplest structures (e.g. I-beams, thin-walled tubes) were determined by *analytical methods* which can only be used for problems with few unknowns. The extension to the solution of more complicated problems was made possible by the development of more general computational techniques. The more recent application of computers caused a revolution not only in structural analysis through the use of the finite element method, but also enabled the development of structural synthesis by numerical methods of mathematical programming.

The complex process of structural design, the *structural synthesis,* consists of the following three phases:

(1) Formulation of the design constraints and the objective function by the design engineer;
(2) Mathematical minimization;
(3) Evaluation of results and their elaboration by the designer.

Optimum design is desirable not only because it enables the lowering of costs by means of mathematical methods, but because it also makes the complex design work more systematic. An economic design which was, until recently, intuitive and routinish, becomes thus more conscious and scientific through the use of structural synthesis. The designer can then take into account all design constraints as well as new research results in order to evaluate the relevant cost function and hence obtain a realistic estimate of the overall cost.

The synthesis results in a higher quality of design and production. However, the definition of actual fabrication constraints and cost functions needs a closer cooperation of designers and technologists.

The main task of the design of structures is to improve *safety* and *economy*. The improvement of safety is governed by analytical checks on problems, such as, for example, *fracture mechanics, stability* and *dynamics*. Economy, on the other hand, is improved by optimum design.

The most efficient means for decreasing the structural weight is the reduction of plate thicknesses. This is limited by the following phenomena:

(1) Limit stress state due to external loads;
(2) Distortions caused by welding;
(3) Buckling due to external loads and residual welding stresses;
(4) Vibration and noise caused by external excitation.

Thus, for the minimum weight design it is necessary to collect the best research results from the field of residual welding stresses, stability and vibration of plated structures and incorporate them into the structural synthesis.

The aim of this book is to introduce the design engineers to the principles and methods of optimum design. Therefore, only comparatively simple structures are investigated. The aim is to concentrate on the basic principles involved, and these can then be applied to more complex systems. Analytical and numerical methods are treated and applied to the optimization of metal structures. In particular, special design problems in welded structures are taken into account by considering the effect of residual welding stresses on the global and local stability, special welding technological constraints, and the costs involved in the welding process.

Part I deals with the general problems of optimum design, the design constraints and cost functions, and gives a brief survey of mathematical methods.

Part II considers the application of optimum design to the following structures: compressed members of welded square box cross-section, plate and box girders, hybrid and sandwich beams loaded in bending and shear, simple planar trusses, frames, and cellular plates.

By using simple formulae derived by analytical methods, many comparisons are made which are very helpful to designers. In the case of two unknowns, the graphical method is used, a technique suitable for simpler optimization problems. Numerical methods of computation such as the SUMT-, the Box-, the backtrack- and the optimum-criteria-methods are also described in detail.

The Appendix contains some special problems of analysis, most of which are based on the results of the author's own research. These include: the calculation of residual stresses due to welding, buckling of compressed struts and plates, static and dynamic response of sandwich beams with outer layers of box cross-section, the analysis of cellular plates, and a Fortran program for the backtrack programming method. Finally, about 400 references as well as a detailed list of symbols are included.

The author takes this opportunity to express his thanks to Professors Otto Halász and Pál Michelberger, corresponding members of the Hungarian Academy of Sciences for reading the text and making valuable suggestions.

The author is also grateful to Dr. Imre Timár, Mech. Eng., Dr. László Szabó, Mech. Eng., Weld. Eng., and Dr. Károly Jármai, Mech. Eng. who helped with the computations of the numerical examples.

The author hopes that this book will prove useful for university students, research workers and practising engineers.

List of Symbols

A	cross-sectional area
A_f	area of a flange
A_T	a quantity characterizing the shrinkage effect caused by welding, Eq. (A1.1)
A_v	cross-section of a weld, Eq. (A1.2)
A_w	area of the web(s)
A_y	area of the plastic zone around a weld, Eq. (A1.7)
a, a_x, a_y	spacings of ribs
a_w	dimension of a fillet weld
B, B_f, B_s	bending stiffnesses of a sandwich beam, Eq. (A4.3)
B_x, B_y	bending stiffnesses of a cellular plate, Eq. (A5.4)
B_{xy}, B_{yx}	torsional stiffnesses, Eq. (A5.4)
B_p	press frame dimension, Fig. 6.13
B_q	shear stiffness of a sandwich beam, Eq. (A4.4)
$B_\omega, B_{\bar{\omega}}$	warping moments, Section 5.2.2.5
b	plate width
b_e	effective plate width
b_f	flange width
b_p	press frame dimension, Fig. 6.13
b_x, b_y	side lengths of a rectangular plate
b_y	average width of the plastic zone around a weld, Eq. (A1.6)
C	curvature, Eq. (A1.4)
C_e, C_f, C_h	constants, Section 7.2.2.1
C_w	deflection constant, Eqs (5.7), (5.53)
c_b, c_L	constants, Eq. (6.3)
c_{fx}, c_{fy}	constants, Eq. (A5.10)
$c_{x\sigma}, c_{y\sigma}$	constants, Eq. (A5.14)
c_m	spring constant
c_N	a constant, Eq. (3.8)
c_0	specific heat, Eq. (A1.1)
c_w	a constant, Eq. (A5.13)
c^*	allowable ratio of deflection

D	diamater
$\mathbf{d}^i$	Eq. (2.4)
E	modulus of elasticity
$E_1 = E/(1 - \nu^2)$	modulus of elasticity for plates
E_{al}, E_s	modulus of elasticity for Al-alloys and steels, respectively
e	distance, eccentricity
F	force
f	eigenfrequency
f_0	prescribed eigenfrequency
G_{al}, G_s	mass of Al-alloy and steel, respectively
G_s, G_d	static and dynamic shear modulus, respectively
G	shear modulus
$\mathbf{G}^i$	Hessian matrix, Eq. (2.3)
g_0	Eq. (A4.12)
g_j	constraint functions
H	torsional stiffness of an orthotropic plate
$\mathbf{H}^i$	approximate Hessian matrix, Eq. (2.8)
h	height
h_j	constraint functions
$\mathbf{I}$	unit matrix
I	arc current, Eq. (A1.2)
I_x, I_y	moments of inertia
I_0	required moment of inertia
I_p	Section 5.2.2.5
I_t	torsional inertia, Section 5.2.2.4
$I_\omega, I_{\overline{\omega}}$	warping section constants, Section 5.2.2.5
i	radius of gyration
$i = \sqrt{-1}$	imaginary unit
K	cost
K_m, K_W, K_p	costs of material, welding and painting, respectively
k_m, k_W, k_p	cost coefficients for material, welding and painting, respectively
k_f, k_w	material cost coefficients for flanges and web, respectively
$k_1, \ldots, k_4$	cost coefficients
k, k_M, k_N, k_τ	buckling factors
L, l	span length, length
L_1	distance of flange splices
M	bending moment
M_p	plastic limit bending moment
M_t	torque
M_u	ultimate bending moment
m	mass
m_{red}	reduced mass
$m = 100M/(R_u h_0^3)$	dimensionless bending moment parameter

m_x, m_y	bending moments per unit length
m_{xy}	twisting moment per unit length
N	normal force
N_u	ultimate normal force
$n = Nh_0/(2M)$	dimensionless compressive force parameter
p	uniform normal load intensity
p	a constant, Eq. (3.6)
Q	shear force
Q_u	limiting shear force
Q_T	specific heat input caused by welding
q	specific shear force
q	specific shear force parameter
q	a constant, Eq. (3.7)
q_0	a constant, Eq. (A1.2)
R_{adm}	admissible stress
R_u	limiting stress
R_e	proportional limit stress
R_{ub}	limiting buckling stress
R_y	yield stress
$r = R_e/R_y$	Eq. (A3.5)
r_k	response factors, Eq. (2.14)
S	statical moment
s^i	Eq. (2.2)
T	temperature
T_0	transmissibility, Eq. (A4.16)
t	thickness
t_0	prescribed minimal plate thickness
t_f, t_w	flange and web thickness, resp.
t_r	rib thickness
$\overline{t}$	time
U	arc voltage, Eq. (A1.2)
V	volume
v	welding speed of travel, Eq. (A1.2)
W, W_x	elastic section modulus
W_p	plastic section modulus
W_0	required section modulus
w	deflection
w_q	deflection due to shear deformation
w^*	allowable deflection
X	shear parameter, Eq. (A4.15)
x_i	variables
Y	Eq. (A4.13)
Y_j	slack variables, Eq. (2.20)
y^i	Eq. (2.10)
z	coordinate

$\alpha = \sqrt{GI_t/(EI_\omega)}$	see Section 5.2.2.5
α	angle of inclination, Fig.5.3
α^i	Eq. (2.4)
$\alpha = a/h$	a parameter, Eq. (7.1)
α_1, α_2	Eq. (A2.2), Table A2.1
α_0	coefficient of thermal expansion
α_0	coefficient, Section 5.2.2.4
α_M, α_B	Eq. (A2.7), Table A2.1
β	limiting elastic plate slenderness for webs, Table A3.1
β_p	limiting plastic plate slenderness for webs, Table A3.1
β_1	a constant, Table A2.1
$\Gamma = b_x/b_y$	Eq. (A5.15), Fig. A5.2
γ	angular distortion
γ_{xy}	angular distortion, Eq. (A5.1)
γ	safety factor
δ	limiting elastic plate slenderness for flanges, Table A3.1
δ_p	limiting plastic plate slenderness for flanges, Table A3.1
$\delta_c = D/t$	diameter/thickness ratio for a circular hollow section
δ_d	log. decrement
ϵ	specific strain
ϵ_y	specific yield strain
ϵ_C	specific strain at centroid, Eq. (A1.4)
ζ_q	Eq. (6.14)
$\eta = b_y/t$	Fig. A1.3
η_b	Eq. (A2.6)
η_d	loss factor
$\eta_h = k_w/k_f$	hybrid beam parameter, Eq. (5.75)
η_0	coefficient of efficiency, Eq. (A1.2)
θ	angular deformation, Section 6.2
θ, θ_d	angles of inclination, Eq. (A3.26)
θ_H, θ_B	Eqs (A5.12), (A5.15)
ϑ	angular-deformation, Section 6.2.3
$\vartheta = b/t$	plate slenderness
$\vartheta = a/t_f$	a parameter, Eq. (7.1)
$\vartheta_p = I_1/I_2$	press frame parameter, Eq. (6.3)
$\vartheta_c = D/h$	a parameter, Eq. (5.60)
λ	column slenderness ratio
$\bar{\lambda} = \lambda/(\pi\sqrt{E/R_y})$	reduced column slenderness ratio
λ^i	optimal step length, Eq. (2.2)
λ_j	Lagrange multipliers, Eq. (2.21)
λ_p	plate slenderness ratio
$\mu = 1 - I_t/I_p$	Section 5.2.2.5
$\mu = t_f/t_r$	a parameter, Eq. (7.1)
μ_q	Eq. (6.14)

μ_0	Section 6.3.1
ν	Poisson's ratio
$\xi = t_2/t_1$	a parameter, Eq. (A5.9)
$\xi_h = R_{yw}/R_{uf}$	hybrid beam parameter, Eq. (5.66)
ρ	density
ρ_q	coefficient of shear stress distribution, Eq. (6.13)
σ_x, σ_y	normal stresses
σ_M	stress due to bending
σ_N	stress due to normal force
σ_c	compressive residual welding stress
$\sigma_\omega, \sigma_{\bar{\omega}}$	normal warping stresses, Section 5.2.2.5
τ	shear stress
τ_{adm}	admissible shear stress
τ_u	limiting shear stress
$\varphi = b/a$	a parameter, Eq. (7.1)
φ_b	coefficient of overall buckling, Eq. (A2.1)
$\varphi_0, \varphi_1, \varphi_2$	angular deformations, Eq. (6.2)
$\varphi_h = h_2/h_1$	a parameter, Eq. (6.20)
χ	sandwich beam parameter, Eq. (A4.2)
$\psi = b_e/b$	effective width ratio
ω	angular frequency
$\omega, \bar{\omega}$	warping coordinates (sectorial areas), Section 5.2.2.5
$\omega_i = I_i/I_0$	ratios of moments of inertia
∇	nabla (vector)

PART I

GENERAL PROBLEMS
IN OPTIMUM DESIGN

Chapter 1

Analytical Phase
of the Structural Synthesis

The three main phases of the structural synthesis are, as mentioned in the preface, as follows:

```
┌─────────────────────────┐
│   Structural synthesis  │
└─────────────────────────┘
```

1. *Analytical phase:* General preparation and the carrying out of the necessary analyses by the designer:
   (a) Selection of materials, profiles, type of structure, production technology;
   (b) Formulation of the design constraints and the cost function

2. *Mathematical phase:* Minimization of the cost function with fulfilment of the design constraints

3. *Phase of evaluation and refinement of the design by the engineer:* Sensitivity analysis, implantation into structural systems, elaboration of design aids

1.1 Selection of Materials, Type of Structure, Production Technology

The application of new materials and/or technologies may result in more economical structures. The most important new materials are steels of improved yield stress and composites.

For example, 37-steels (tensile strength 370 MPa) can be replaced by 52-steels (tensile strength 520 MPa, yield stress 340 − 400 MPa). Steels considered in the European Recommendations (1978) are as follows: Fe 360, 430 and 510 with yield stress of 235, 275 and 355 MPa, respectively. In some countries other weldable high strength structural steels are developed as well, e.g. steels with yield stress of 390, 410, 440 and 590 in USSR (SNiP 1982), HW 70, 80, 90 steels in Japan, A 514 steel (yield stress 690 MPa) in USA, etc. More detailed information about the high strength steels used in steel construction can be found, for example, in the works of Zherbin (1974), and Kato and Okumura (1976).

The use of high strength steels is economical mainly in those structural parts subjected to static tension (storage tanks, pressure vessels, trusses) or in members the design of which is governed by compression in the plastic buckling region (columns, trusses). On the other hand, in structures with active displacement constraints (radiotelescopes, machine structures) or those where compression in the elastic buckling region is the governing criterion (slender bars) it is not economical to increase the yield stress. High strength steels are also less economical in structures loaded with pulsating loads, since the fatigue strength of welded joints does not increase in proportion to the yield stress, even though some new techniques for improving the fatigue strength of welded joints are now available (e.g. Shimada *et al.* 1978). Furthermore, one should note that, in structures subjected to bending, hybrid I-sections, with flanges made of steel of higher strength than the web, result in a more economical solution than that where the cross-section is kept homogeneous (see Section 5.2.8).

As to the types of modern structures which have proved to be especially economical one can mention: tubular trusses, prestressed metal beams, shells, and pneumatic, sandwich and cable structures.

In the field of modern production technology recent developments include: new welding methods (electron-beam, laser), extrusion of aluminium-profiles, cold-forming of rectangular tubes, etc.

Finally, one must emphasize that the minimum cost design of structures of new type is difficult, because cost data are not yet readily available.

1.2 Design Constraints

There are three types of design constraints: constraints on strength, those on production, and any other requirements (such as, for example, aesthetical considerations).

Strength constraints may be formulated on the basis of the possible failures or other limit states which would render the structure unfit for use.

Production constraints, on the other hand, are mainly related to the actual dimension of a structure (e.g. available thicknesses).

Constraints should be formulated as *inequalities,* because it is not known in advance whether a constraint is active or not.

Design requirements are formulated in rules, codes or standards, which are, unfortunately, different in various countries, and which are also dependent on the actual type of structure. Therefore, in numerical optimum design it is impossible to give optimum data generally valid for all structures. However, data published in this book give, nevertheless, a good starting point on optima, because the differences between the various prescriptions are not very significant and, also, because objective functions are not very sensitive near the optimum. Therefore, on the basis of principles and methods treated in this book, the designer should be able to formulate and solve his own problems according to actually valid prescriptions.

1.2.1 Strength Constraints

According to the formulation of the European Recommendations (1978) there are two categories of limit states:

(a) Ultimate limit states:
    (1) Loss of equilibrium;
    (2) Rupture of critical section(s) of the structure or excessive deformation;
    (3) Transformation of the structure into a mechanism;
    (4) General or local instability;
    (5) Fatigue;
(b) Serviceability limit states:
    (1) Deformations adversely affecting the appearance or efficiency of the structure (e.g. a deforming roof structure can result in an increase in loading due to ponding);
    (2) Local damages (cracking, localized buckling, splitting, etc.);
    (3) Vibrations (due to wind or machinery) adversely affecting the comfort of users or the use of the structure.

The verifications of ultimate limit states can be made by means of elastic or plastic design methods; those of serviceability limit states must be carried out in the elastic range.

The verifications should be made with factored loads, i.e. with representative values of the various actions, calculated by multiplying the permanent and variable loads by safety (γ_{perm}, γ_{var}) and simultaneity (ψ_s) factors, as follows:

$$\gamma_{perm} F_{perm} + \gamma_{var}\left(F_{var\,max} + \sum_{i \neq max} \psi_{si} F_{var\,i}\right).$$

Different combinations of actions should be calculated, including those components which might arise from accidental causes.

(a) *Verifications concerning the ultimate limit states,* using elastic design methods, are as follows:

(1) Constraint on the maximum stress. In the case of a static plane stress state, using the Mises' formula one obtains

$$\sqrt{\sigma_x^2 + \sigma_y^2 - \sigma_x\,\sigma_y + 3\tau^2} \leqslant R_u$$

where R_u is the ultimate limit stress. For fillet welds, and according to the Design Rules (1976), the following formula may be used:

$$\beta\sqrt{\sigma_\perp^2 + 3(\tau_\perp^2 + \tau_\parallel^2)} \leqslant R_u$$

where $\beta = 0.70$ for steel Fe 360, and $\beta = 0.85$ for Fe 510 ($\sigma_\perp, \tau_\perp$ and $\tau_\parallel$ are stress components perpendicular and parallel to the axis of the weld, respectively).

(2) Constraint due to overall stability (buckling of compressed struts, stiffened plates and shells, buckling of beam-columns subjected to compression and biaxial bending, lateral buckling).

(3) Constraint due to local stability (buckling of plate elements, local buckling of shells).

(4) Constraint due to fatigue: in the case of time-varying loads, R_u or R_{adm} (admissible stress) should be determined according to stress concentration (fatigue group), number of cycles, load spectrum and environment. For special structures (e.g. stiffened plates or shells used in aircraft structures) a constraint due to fatigue crack growth arrest can be used (Dobbs and Nelson 1978). A special problem is the corrosion fatigue (as, for example, in cables) or low cycle fatigue of irradiated structural parts of nuclear reactors.

(b) *Verifications concerning the serviceability limit states*

(1) *Constraint imposed by static deformations* may be dominant mainly in the case of machine structures (presses, cranes, radiotelescopes, TV-antennae, etc.). Deflection constraints may be written in the form

$$w_{max} \leqslant c^* L$$

where L is the length of the element. The constant c^* varies between $1/100$ and $1/300$ in the case of buildings; for cranes $c^* = 1/500 - 1/1000$, while for press frames $c^* = 1/800 - 1/10000$ (see, e.g. (Morgenstern 1975)). The deflection constraint may also be relevant for structures made of Al-alloys (see Section 5.2.2.2).

(2) *Constraints due to vibration.* The minimum value for the first natural frequency can be prescribed; e.g., $f_1 \geqslant 3.5$ Hz for floor slabs (Verner 1976). The maximum amplitude of vibration can also be restricted; e.g., for a flame cutting machine $a_{max} \leqslant 0.14$ mm (Klinnert and Möbius 1975). For instance, the paper of Cheng and Botkin (1976) solves the minimum weight design of frames with constraints imposed by frequency and dynamic stress and displacement considerations, taking also into account the damping characteristics of the system.

The constraint on vibration damping may be formulated by prescribing the time during which the initial maximum amplitude w_{max} decreases to an allowable value w^*. The differential equation of natural damped vibration of a one-mass system is

$$m\ddot{w} + (1 + i\eta_d)w/c_m = 0$$

where c_m is the spring constant, $\eta_d = \delta_d/\pi$ is the loss factor, $\delta_d = $ log. decrement. Solving the equation one obtains

$$w_{max}/w^* = \exp(\omega\eta_d\overline{t}/2)\;;\qquad \omega = (m\,c_m)^{-1/2}\;.$$

The damping constraint is

$$\overline{t} = \frac{\ln(w_{max}/w^*)}{f\delta_d} \leqslant \overline{t}^{\,*}$$

where $f = \omega/(2\pi)$ [Hz] eigenfrequency, $w^* = 0.5$ mm (Boguslavskiy 1961), $\overline{t}^{\,*} = $ $= 12 - 15$ s (ČSN 270103), (Demokritov 1962). $\delta_d = 0.07$ for crane box girders of height $h = L/18 - L/20$, while $\delta_d = 0.05$ for $h < L/20$ (L = span length). w_{max} can be calculated as the maximum static deflection.

1.2.2 Technological Constraints

In the case of welded structures it is necessary to prescribe the minimum plate thickness suitable for the applied welding technology. These thickness limits are given in Table 1.1 (Mathisen 1976). Limits for other size parameters such as the width or height of a beam may also be prescribed.

Table. 1.1
Plate Thickness Limits for Various Welding Processes for Steels

Welding process	Plate thickness [mm] for single-pass welding	multipass welding
Ultrasonic	1	
Microplasma	0.1 – 2	
Laser 2 kW	3	
Resistance spot & seam	0.1 – 5	
Plasma	0.3 – 6	
Gas tungsten arc (TIG)	4	4 – 10
Gas	0.5 – 4	4 – 10
Laser 20 kW	0.5 – 15	
MIG dip transfer	0.5 – 4	4 – 15
MIG	2 – 10	10 – 50
Manual-metal-arc	2 – 6	6 – 150
Electron-beam 5 kW	30	
Electron-beam 25 kW	1 – 60	
Electron-beam 75 kW	5 – 300	
Submerged arc	5 – 25	25 – 250
Electroslag	25 – 400	

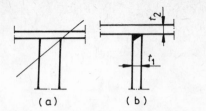

Fig. 1.1 – Thickness ratios of a butt welded T-joint: (a) incorrect; (b) correct

A special welding requirement relates to the thickness ratio of a T-joint (Fig. 1.1). To avoid the inadequate joint shown in Fig. 1.1(a) the condition

$$t_2 \geqslant \alpha_w t_1 \, , \qquad \alpha_w = 0.5 - 1$$

must be fulfilled.

1.3 Objective Functions

As objective/merit functions only mass- and cost functions are treated here. The latter should be as realistic and detailed as possible. Unfortunately, the costs vary in time and depend on many factors. It is sufficient to consider only those costs which significantly affect the variables to be optimized. Many authors take into consideration only the mass-function (minimum weight design). Such approach gives, in many instances, suitable data for designers. In other cases, however (e.g. in the optimum design of welded stiffened plates), the costs of a structure optimized with regard to minimum mass and minimum cost respectively, may differ considerably (see e.g. Chapter 7 or the article of Caldwell and Hewitt (1976)).

In the optimum design of a homogeneous, unstiffened girder with constant cross-section it is sufficient to minimize the cross-sectional area. In the case of hybrid beams (see Section 5.2.8) the different steel prizes of web and flanges should also be taken into account.

The cost of a welded structure consists of material and labour costs. The labour costs are as follows: costs of cutting, joint preparations, tacking, welding, chipping, heat treatment, non-destructive testing, handling, erection, protection against corrosion (painting) (Quinn 1977).

The costs of welding consist of cost of electrodes, gas, current, machine, labour and inspection (Czesany 1972), (Herden 1967). The welding costs depend on plate thickness and on type of technology used. Some estimated shop hours for manual arc welding are given in the European Recommendations (1978). Detailed cost analysis of ship structures can be found in the work of Moe and Lund (1968). The minimum cost design of steel structures for industrial buildings has been studied by Lee and Knapton (1974), Lipson and Russell (1971), Thomas and Brown (1977).

Van Douwen (1981) has pointed out that the cost of labour increases much more faster than the material cost, therefore designers have to consider the labour costs in the optimization. For example, in the selection of the type of joints (beam-to-column connections) the fabrication cost should be considered in the first place.

In the third phase of structural synthesis the *sensitivity analysis* plays an important role, because it characterizes the objective function behaviour in the region of the optimum. The simplest method of sensitivity analysis is to give the measure of increasing of the objective function relating to the variation of the variables in a certain interval (e.g. ±10%). On the basis of this information, the designer can determine the final (e.g. rounded) values of variables which do not increase the cost significantly.

In most cases of optimum design of metal structures the objective functions are not very sensitive in the vicinity of the optimum. Numerical examples show that the more complex the cost function the smaller is its sensitivity (refer, for example, to Chapters 5 and 7).

The optimum design of *structural systems* (e.g. steel structures for industrial buildings) is a complex problem. In the *decomposition* of the whole problem, i. e. in *suboptimization* of structural parts, it should be born in mind that these elements are components of the whole system.

In the case of *additional objective functions* special mathematical methods should be applied (see, for example, the works of Peschel and Riedel (1975) and Gerasimov and Repko (1978)).

Attention should be given to the *dangers of optimum design.* If several different stability criteria are fulfilled the load-carrying capacity of the optimum structure may be reduced significantly by the interaction of these different instability modes. For instance, in the case of optimum design of compressed thin-walled struts (see Chapter 3) or that of compressed stiffened plates, the local buckling of the plate elements may decrease the overall buckling strength.

Chapter 2

Mathematical Methods for Structural Synthesis

2.1 Survey of Methods

In optimum design procedure the minimum of the objective function is searched which fulfils the inequality and equality constraints as follows:

$$
\left.
\begin{aligned}
&\min_{\mathbf{x}} f(\mathbf{x}) & &\mathbf{x} = (x_1, \ldots, x_i, \ldots, x_n)^T \\
&g_j(\mathbf{x}) \geqslant 0 & &j = 1, 2, \ldots, m \\
&h_j(\mathbf{x}) = 0 & &j = m + 1, \ldots, p
\end{aligned}
\right\}
\qquad (2.1)
$$

The unknowns (n in number) may be: (1) structural sizes (thicknesses, widths of plates); (2) geometric data (coordinates of nodes of a truss, location of ribs in a stiffened plate); (3) material quality.

The functions may be continuous or the unknowns may be defined by series of discrete values (e.g. fabricated plate thicknesses).

It is often useful to transform the variables or replace them by dimensionless ones. From computational aspects it is advisable to scale the variables.

The constraints (p in number) were treated in Section 1.2. Constraints may be explicit or implicit. In many cases it is required that the variables be positive ($x_i \geqslant 0$), or their upper and lower limit may be prescribed ($x_i^L \leqslant x_i \leqslant x_i^U$). The functions f, g, h may be linear or nonlinear. The number of variables may be small ($n < 20$) or large (e.g., spatial trusses or structures partitioned to finite elements may result in hundreds of variables).

Figure 2.1 shows some examples of structures with small numbers of variables which are treated in this book. These are: (a) compressed strut of square box cross-section (Chapter 3); (b) welded I-beam subjected to bending and shear (Chapter 5); (c) cellular plate (Chapter 7); (d) closed frame with rods of welded box cross-section (Chapter 6).

In structural synthesis problems it is characteristic that the number of constraints is larger than that of variables ($p \geqslant n$).

Problems with *two variables* may be treated graphically. Plotting the limit curves $g_j(x_1, x_2) = 0$ in the coordinate-system $x_1 - x_2$ we get the *feasible region*

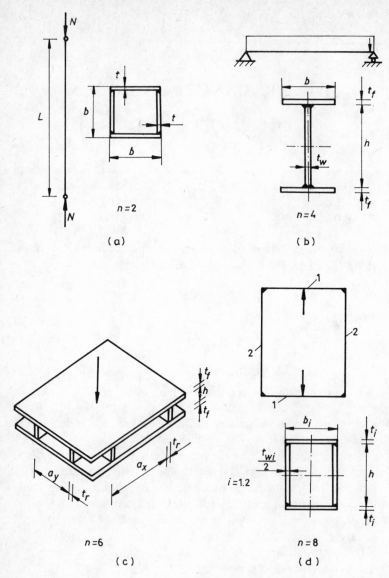

Fig. 2.1 − Some examples of structures with a small number of variables

(Fig. 2.2). The contours of the objective function $f(x_1, x_2) = C_k$ ($k = 1, 2, ..., s$) may also be drawn, one of them touching the feasible region at the optimum point. If the feasible region is concave (Fig. 2.3) (e.g. stress constraints in statically indeterminate rod structures) it is only a local optimum. The mathematical condition for a local optimum was formulated by the *Kuhn − Tucker's optimality criteria* (see Section 2.4).

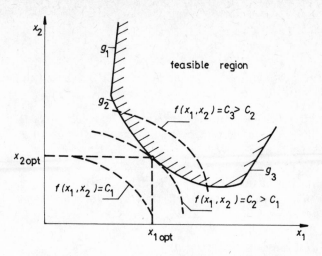

Fig. 2.2 – Graphical determination of the optimum solution in a two-dimensional space
in the case of a convex feasible region

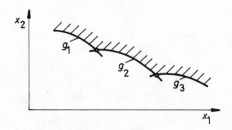

Fig. 2.3 – Local optima in the case of a concave feasible region

The mathematical methods of optimum design can be divided into two groups: *analytical* and *numerical* methods.

Analytical methods include differentiation, calculus of variations (see, e.g. (Niordson and Olhoff 1979), (Gol'dstein and Solomeshch 1980)), the method of Lagrange-multipliers (see, e.g. (Kicher 1966)). Closed form solutions for optimum problems can be obtained only for the simplest cases. For rods of variable cross-section functions can be derived for the optimum shape (see, e.g., (Banichuk 1980)). However, since it is difficult to fabricate these shapes, it is advisable to replace them by rods of piecewise constant cross-section. Such rods can then be analysed by means of the finite element method.

The optimality-criteria-methods are analytical—numerical, because they combine the Lagrange-multipliers and Kuhn-Tucker-criteria with the finite element method (see Section 2.4).

Numerical, i.e. *mathematical programming methods* can be divided in *linear* and *nonlinear* groups.
Nonlinear programming methods are as follows.

(1) *One-dimensional minimization methods:* interval halving, Fibonacci, golden section, quadratic or cubic interpolation, direct root method.

(2) *Multivariable unconstrained methods*
    (a) *Methods using derivatives* (gradient methods)
        Steepest descent, conjugate gradient (Fletcher—Reeves)
        Newton
        Variable metric (Davidon — Fletcher — Powell)
    (b) *Methods without using derivatives* (direct search methods)
        Random search
        Pattern search (Hooke—Jeeves, Powell)
        Rosenbrock (unconstrained)
        Nelder — Mead (simplex)

(3) *Multivariable constrained methods*
    (a) *Methods using derivatives*
        Zoutendijk's method of feasible directions
        Rosen's gradient projection method (used only for linear constraints)
        Penalty function methods (SUMT)
    (b) *Methods without using derivatives*
        Complex (Box)
        Rosenbrock (constrained)

(4) *Other methods*
        Geometric programming
        Dynamic programming
        Stochastic programming
        Theory of games
        Optimal control theory
        Combinatorial methods: branch and bound, backtrack.

These mathematical methods are treated in many books, e.g. (Hupfer 1970), (Leśniak 1970), (Fox 1971), (Himmelblau 1971), (Sergeev and Bogatyrev 1971), (Kreko 1974), (Gallagher and Zienkiewicz 1973), (Majid 1974), (Batishchev 1975), (Reitman and Shapiro 1976), (Haug and Arora 1978), (Rao 1978), (Iyengar and Gupta 1980), (Fletcher 1980—81), (Malkov and Ugodchikov 1981), (Ol'khov 1981), (Kirsch 1982).

In the book of Kuester and Mize (1973) Fortran programs are listed for the following methods: simplex (linear programming), zero-one programming (linear integer programming), geometric programming, dynamic programming, Fibonacci, Nelder—Mead, Hooke—Jeeves, Rosenbrock (unconstrained and constrained), Powell, Fletcher—Reeves, Box, Rosen, SUMT + variable metric.

In the present work we describe in detail only some of the methods which are predominantly applied to structural optimization and which are also used in the author's research investigations. These methods are: Davidon — Fletcher — Powell, SUMT, optimality-criteria, Nelder — Mead, Box, backtrack.

2.2 The Davidon — Fletcher — Powell Method

Since the gradient vector $\nabla f(\mathbf{x})$ denotes the direction of steepest ascent, i.e. that in which the function increases at the fastest rate, the minimum should be searched in the direction of negative gradient (direction of steepest descent). Using the *steepest descent method,* an iterative search procedure should be performed, in which, after the ith approximation, we take a new point

$$\mathbf{x}^{i+1} = \mathbf{x}^i - \lambda^i \nabla f(\mathbf{x}^i) \tag{2.2}$$

where λ^i is the optimum step length (in the direction of steepest descent) which minimizes the function

$$f(\mathbf{x}^i + \lambda^i \mathbf{s}^i); \qquad \mathbf{s}^i = \nabla f(\mathbf{x}^i) \,.$$

The convergence of the steepest descent method is slow, and thus numerous modifications have been worked out to improve it; these include the *conjugate gradient method* of Fletcher — Reeves and the *variable metric method* of Davidon — Fletcher — Powell.

The quadratic approximation of $f(\mathbf{x})$, i.e. the three terms of the Taylor's series expansion of $f(\mathbf{x})$ about the point $\mathbf{x}^i$, gives

$$f(\mathbf{x}) = f(\mathbf{x}^i) + \nabla f(\mathbf{x}^i)^T (\mathbf{x} - \mathbf{x}^i) + \frac{1}{2}(\mathbf{x} - \mathbf{x}^i)^T \mathbf{G}^i (\mathbf{x} - \mathbf{x}^i) \tag{2.3}$$

where

$$f(\mathbf{x}^i)^T = \left[\frac{\partial f}{\partial x_1}, \frac{\partial f}{\partial x_2}, \cdots, \frac{\partial f}{\partial x_n} \right].$$

$\mathbf{G}^i = \nabla^2 f(\mathbf{x}^i)$ is the Hessian matrix. In *Newton's method* the step difference

$$\mathbf{d}^i = \alpha^i \mathbf{s}^i = \mathbf{x}^{i+1} - \mathbf{x}^i \tag{2.4}$$

is obtained by differentiating

$$f(\mathbf{x}^{i+1}) = f(\mathbf{x}^i) + \nabla f(\mathbf{x}^i)^T \mathbf{d}^i + \frac{1}{2}\mathbf{d}^{iT} \mathbf{G}^i \mathbf{d}^i \tag{2.5}$$

with respect to $\mathbf{d}^i$ and equating the resulting expressions to zero to give

$$\mathbf{d}^i = -(\mathbf{G}^i)^{-1} \nabla f(\mathbf{x}^i) \,. \tag{2.6}$$

Thus, Newton's method requires the inverse of the Hessian matrix.

The quasi-Newton methods use an approximation of the inverse matrix. The iterative procedure of the *Davidon — Fletcher — Powell-method* consists of the following calculations:

$$x^{i+1} = x^i + \lambda_{\min}s^i = x^i + \alpha^i s^i = x^i + d^i \tag{2.7}$$

where

$$s^i = -H^i \nabla f(x^i). \tag{2.8}$$

H^i is an approximate matrix replacing $(G^i)^{-1}$. In the first step H^0 is usually taken as the identity matrix $H^0 = I$. In the next iteration we use the following formula for H:

$$H^{i+1} = H^i + \frac{d^i d^{iT}}{y^{iT} d^i} - \frac{H^i y^i y^{iT} H^i}{y^{iT} H^i y^i} \tag{2.9}$$

where

$$y^i = \nabla f(x^{i+1}) - \nabla f(x^i). \tag{2.10}$$

Broyden has proposed an alternative recursion relation

$$H^{i+1} = H^i + \frac{(s^i - H^i y^i)(s^i - H^i y^i)^T}{(s^i - H^i y^i)^T y^i}. \tag{2.11}$$

A survey of recursion relations is given, for example, by Himmelblau (1971).

The flow chart of the DFP-method appears in Fig. 2.4. For the computation of $\lambda_{\min}$ the one-dimensional minimization methods may be used. A convergence criterion is

$$\nabla f(x^i)^T (G^i)^{-1} \nabla f(x^i) < \epsilon \tag{2.12}$$

where ϵ is usually taken as 10^{-6}. Alternatively, another criterion may be

$$\nabla f(x^i) < \epsilon. \tag{2.13}$$

To illustrate the DFP-method consider the following *numerical example* where the function

$$f(x_1, x_2) = 4(x_1 - 5)^2 + (x_2 - 6)^2 + 3$$

is to be minimized. Let the starting point be $x^0 = [6;7]^T$ and $\epsilon = 10^{-6}$. The gradient vector is

$$\nabla f = [8(x_1 - 5); \ 2(x_2 - 6)]^T$$

which at the point x^0 becomes $[8; 2]^T$. The Hessian matrix and its inverse are then

$$G = \nabla^2 f = \begin{bmatrix} 8 & 0 \\ 0 & 2 \end{bmatrix}; \qquad G^{-1} = \begin{bmatrix} 1/8 & 0 \\ 0 & 1/2 \end{bmatrix}.$$

Since $\nabla f^{0T} G^{-1} \nabla f^0 = 10 > \epsilon$, the convergence criterion is not fulfilled. We next compute the direction vector as

$$s^0 = -H^0 \nabla f^0 = -\nabla f^0 = [-8; -2]^T.$$

Since

$$f(x^0 + \lambda s^0) = 4(1 - 8\lambda)^2 + (1 - 2\lambda)^2 + 3$$

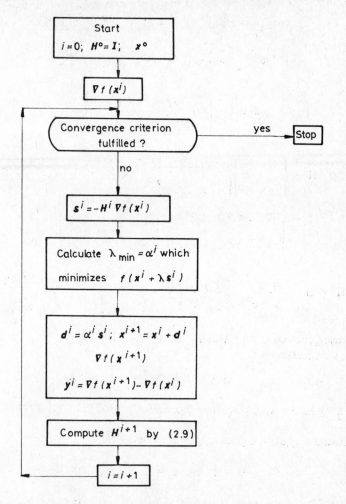

Fig. 2.4 — Flow chart of the Davidon — Fletcher — Powell method

is a simple function of λ, analytical differentiation may be used instead of a numerical one-dimensional search: the condition $df/d\lambda = 0$ then gives $\lambda_{min} = \alpha^0 = 0.13077$.

$$\mathbf{d}^0 = \alpha^0 \mathbf{s}^0 = [-1.04616; \quad -0.26154]^T$$

and the new point is

$$\mathbf{x}^1 = \mathbf{x}^0 + \alpha^0 \mathbf{s}^0 = [4.95384; \quad 6.73846]^T$$

where the gradient is

$$\nabla f^1 = [-0.36928; \quad 1.47692]^T.$$

According to (2.9) and (2.10)

3*

$$y^0 = \nabla f^1 - \nabla f^0 = [-8.36928; \ -0.52308]^T$$

$$y^{0T} d^0 = 8.89241; \quad y^{0T} H^0 y^0 = 70.31846;$$

$$d^0 d^{0T} = \begin{bmatrix} 1.09445 & 0.27361 \\ 0.27361 & 0.06840 \end{bmatrix}$$

$$H^0 y^0 y^{0T} H^0 = \begin{bmatrix} 1 & 0 \\ 0 & 1 \end{bmatrix} \begin{bmatrix} -8.36928 \\ -0.52308 \end{bmatrix} [-8.36928; \ -0.52308] \begin{bmatrix} 1 & 0 \\ 0 & 1 \end{bmatrix} =$$

$$= \begin{bmatrix} 70.04485 & 4.37780 \\ 4.37780 & 0.27361 \end{bmatrix}$$

$$H^1 = H^0 + \frac{d^0 d^{0T}}{y^{0T} d^0} - \frac{H^0 y^0 y^{0T} H^0}{y^{0T} H^0 y^0} = \begin{bmatrix} 0.12695 & -0.03149 \\ -0.03149 & 1.00380 \end{bmatrix}.$$

The convergence criterion yields $\nabla f^1 G^{-1} \nabla f^1 = 1.1077 > \epsilon$, so that further calculations are necessary:

$$s^1 = -H^1 \nabla f^1 = [0.09339; \ -1.49416]^T;$$

$$f(x^1 + \lambda s^1) = 4(-0.04616 + 0.09339\lambda)^2 + (0.73846 - 1.49416\lambda)^2 + 3.$$

The condition $df/d\lambda = 0$ gives $\lambda_{min} = \alpha^1 = 0.49423$;

$$x^{(2)} = x^1 + \alpha^1 s^1 = [5.0000; \ 6.0000]^T.$$

This is the optimum point, because $f^{(2)} = [0; 0]^T$ and the convergence criterion is thus fulfilled.

2.3 The SUMT Method

The SUMT (sequential unconstrained minimization technique) method was developed by Fiacco and McCormick (1968) on the basis of Carroll's (1961) proposal. The minimization problem (2.1) is converted into a sequence of unconstrained problems by defining the *P*-function as follows:

$$P(x, r_k) = f(x) + r_k \sum_{j=1}^{m} \frac{1}{g_j(x)} + r_k^{-1/2} \sum_{j=m+1}^{p} h_j^2(x) \qquad (2.14)$$

or, in another version

$$P(\mathbf{x}, r_k) = f(\mathbf{x}) - r_k \sum_{j=1}^{m} \ln g_j(\mathbf{x}) + r_k^{-1} \sum_{j=m+1}^{p} \min[0, \, h_j(\mathbf{x})]^2. \quad (2.15)$$

The second and third terms are the *penalty functions*. This P-function is repeatedly minimized for a sequence of decreasing values of r_k. The weighting (or "response") factors r_k are positive and form a monotonically decreasing sequence of values:

$$r_1 > r_2 > \ldots > 0; \quad r_{k+1} = r_k/c; \quad c > 1 .$$

The proof of the convergence requirement for the method, i.e.

$$\lim_{r_k \to 0}[\min P(\mathbf{x}, r_k)] = \min f(\mathbf{x}) \qquad (2.16)$$

is described by Fiacco and McCormick (1968), and also by Rao (1978).

The flow diagram of the SUMT method for the version (2.15) without equality constraints (interior penalty function method) is shown in Fig. 2.5.

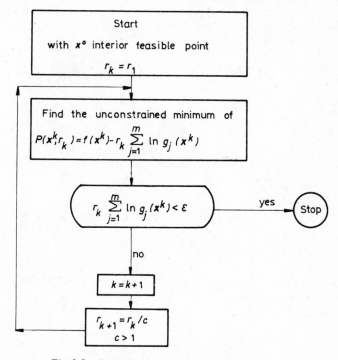

Fig. 2.5 – Flow chart of the SUMT method

Kavlie and Moe (1971) have used a modified version for cases without equality constraints:

$$P(\mathbf{x}, r_k) = f(\mathbf{x}) + r_k \sum_{j=1}^{m} \frac{1}{g_j(\mathbf{x})} ; \quad \text{for} \quad g_j(\mathbf{x}) \geqslant \epsilon \qquad (2.17a)$$

$$P(\mathbf{x}, r_k) = f(\mathbf{x}) + r_k \sum_{j=1}^{m} \frac{2\epsilon - g_j(\mathbf{x})}{\epsilon^2} ; \quad \text{for} \quad g_j(\mathbf{x}) < \epsilon . \quad (2.17b)$$

The following simple *example* illustrates the SUMT method.

$$f(x) = ax$$

subject to

$$g(x) = x - b \geqslant 0 .$$

Using (2.14) we get

$$P(x, r_k) = ax + r_k \frac{1}{x - b} .$$

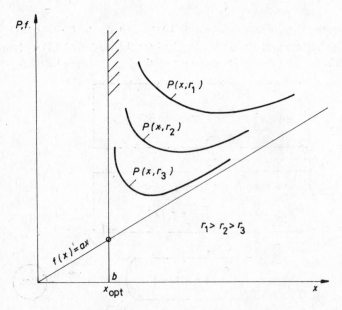

Fig. 2.6 – A simple example of the SUMT method

Figure 2.6 shows that the minima relating to $r_1 > r_2 > r_3$ converge to the minimum of $f(x)$. This can also be proved by calculation. Condition $dP/dx = 0$ gives

$$x_{\min} = b + \sqrt{r_k/a}$$

$$P(x_{\min}, r_k) = ab + 2\sqrt{ar_k}$$

and

$$\lim_{r_k \to 0} P(x_{\min}, r_k) = ab = \min f(x).$$

Schmit and coworkers have developed an advanced automated design procedure (ACCESS –Approximation Concepts Code for Efficient Structural Synthesis) for large structural systems (Schmit and Miura 1976, 1978). ACCESS 1 can handle problems with up to 300 finite elements, 300 displacement degrees of freedom, 120 independent design variables and 5 distinct load conditions. In ACCESS 2 thermal effects, fiber composite materials, and natural frequency constraints are introduced in addition to the usual stress, deflection and member size limitations. Design variable linking, constraint deletion, and explicit constraint approximation are used to effectively combine finite element and nonlinear mathematical programming techniques. In the optimization procedure the NEWSUMT method is used with the following penalty functions (Cassis 1974), Haftka and Starnes 1975), (Cassis and Schmit 1976a):

$$
\left.
\begin{aligned}
P(\mathbf{x}, r_k) &= f(\mathbf{x}) + r_k \sum_{j=1}^{m} \frac{1}{g_j(\mathbf{x})} && \text{for} \quad g_j(\mathbf{x}) \geqslant \epsilon \\
P(\mathbf{x}, r_k) &= f(\mathbf{x}) + r_k \sum_{j=1}^{m} \frac{1}{\epsilon}\left[\frac{1}{\epsilon^2} g_j^2(\mathbf{x}) - \frac{3}{\epsilon} g_j(\mathbf{x}) + 3\right] && \text{for} \quad g_j(\mathbf{x}) < \epsilon
\end{aligned}
\right\}
\quad (2.18)
$$

Prasad and Haftka (1979a) use a cubic penalty function. Fleury (1979) has developed a new method combining ACCESS with the optimality criteria method. In ACCESS 3 approximation concepts and dual methods are combined (Schmit and Fleury 1980). A detailed history of the development of these methods is described by Schmit (1981).

Some of the applications of the SUMT-method include: minimum weight design of continuous girders and frames of piecewise constant thin-walled I-, U- and box cross-section (Nelson and Felton 1972); minimum weight design of welded I- and box girders (Akita and Kitamura 1972); minimum weight design of frames with frequency constraints (Cassis 1974), (Shamie and Schmit 1975); shape optimization of plates and pressure vessel bottoms combining SUMT with the finite element method (Ramakrishnan and Francavilla 1974–75); optimal design of a vertically corrugated bulkhead (4–6 variables) which has resulted in 15% weight savings compared with previous structures of that type (Kavlie et al. 1966); minimization of the midship section of OBO-carriers (8 variables) which has given approx. 2% savings in the total hull weight (Moe 1974); minimum cost design of decks of a car-carrier ship resulting in 10% savings in the total cost of production (Moe 1969); minimum weight design of compressed struts of welded square box cross-section (Farkas 1977e); minimum weight design of welded I- and box beams subjected to bending and shear (Farkas 1978c); minimum cost design of sandwich beams with thin faces (Tímár 1981); optimum design of plates subjected to bending using SUMT together with the finite element method (Prasad and Haftka 1979b); minimum weight design of frames with reliability constraints in the case of earthquake loadings (Davidson et al. 1980); minimum weight design of frames with discrete variables (Liebman et al. 1981).

2.4 Optimality Criteria Methods

Consider the problem

$$\text{mi } f(\mathbf{x}) \qquad x = (x_1, \ldots, x_i, \ldots, x_n)^T$$

subject to   (2.19)

$$g_j(\mathbf{x}) \leqslant 0 \qquad j = 1, \ldots, p$$

in which the Lagrange multipliers λ_j are used. Inequality constraints may be converted to equalities by introducing the slack variables Y_j:

$$g_j(\mathbf{x}) + Y_j^2 = 0. \tag{2.20}$$

The Lagrangian can then be written as

$$L(\mathbf{x}, \lambda_j, Y_j) = f(\mathbf{x}) + \sum_{j=1}^{p} \lambda_j [g_j(\mathbf{x}) + Y_j^2]. \tag{2.21}$$

Criteria for a local minimum of L are as follows:

$$\frac{\partial L}{\partial \mathbf{x}} = \nabla f(\mathbf{x}) + \sum_{j=1}^{p} \lambda_j \nabla g_j(\mathbf{x}) = 0 \tag{2.22}$$

$$\frac{\partial L}{\partial \lambda_j} = g_j(\mathbf{x}) + Y_j^2 = 0 \tag{2.23}$$

$$\frac{\partial L}{\partial Y_j} = 2\lambda_j Y_j = 0. \tag{2.24}$$

If a *constraint is active* (i.e. $g_j = 0$) $Y_j = 0$ and $\lambda_j \geqslant 0$ according to (2.23) and (2.24). On the other hand, for a *passive constraint* $g_j < 0$, $Y_j \neq 0$ and $\lambda_j = 0$. In further calculations, instead of (2.23) and (2.24). we use the criteria

$$\lambda_j \geqslant 0 \quad \text{and} \quad \lambda_j g_j = 0. \tag{2.25}$$

The criteria (2.22) and (2.25), i.e.

$$\nabla f(\mathbf{x}) = -\sum_{j=1}^{p} \lambda_j \nabla g_j(\mathbf{x})$$

$$\lambda_j \geqslant 0; \qquad \lambda_j g_j = 0 \tag{2.26}$$

are the *Kuhn –Tucker optimality criteria*. The first condition states that the necessary criterion for a local optimum is that the gradient to the objective function should be capable of being expressed as the negative of a linear combination of the constraint gradients; in graphical form this results in the gradient of the objective function being within the cone of the constraint-surface normals (Fig. 2.7 (a)).

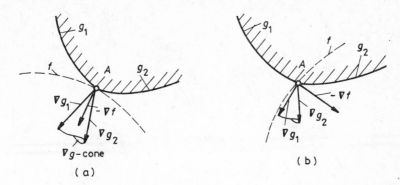

Fig. 2.7 — Graphical representation of the Kuhn–Tucker optimality criterion.
(a) Point A is the optimum; (b) point A is not an optimum

If all constraints are active, the conditions (2.22) and the equations $g_j = 0$ give $n + p$ equalities for $n + p$ unknowns of x_i and λ_j. Numerous solution methods have been proposed, because the solution depends on the type of objective function and constraints. Optimization of rod structures has been the subject of main interest so far, although plates and shells have also been considered.

A detailed description for the solution of trusses and frames is given in Chapters 3 and 6, respectively.

2.5 The Nelder — Mead Method

Among the direct search methods, the Box (or "complex") method is the most suitable technique for nonlinear function minimization problems with nonlinear inequality constraints. Since the Box method is based on the Nelder — Mead method, the latter will first be discussed here.

The Nelder — Mead (1965) method is also known as the *simplex method* (not be confused with the simplex method for linear programming). *Simplex* is a geometric figure formed by a set of $m \geqslant n + 1$ points in n-dimensional space. A simplex is formed by addition of some vectors to the first starting point vector x_{i1} $(i = 1, \ldots, n)$; $x_{ik} = x_{i1} + a_{ik}$ $(k = 2, \ldots, n + 1)$. Figure 2.8 shows a simplex of $m = 4$ points in the two-dimensional space.

The simplex method consists of iterative movements of the simplex towards the optimum point. This method was originally developed by Spendley *et al.* (1962). The movement of the simplex is achieved by using reflexion, expansion and contraction.

The objective function values are calculated at each point of the simplex. The point of the worst (maximum) objective function value (x_{ik}^{Worst}) is rejected and a

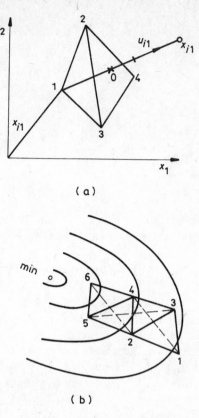

(a)

(b)

Fig. 2.8 – (a) A simplex of $1 - 2 - 3 - 4$ points; x_{i0} centroid of points $2 - 3 - 4$ remained after eliminating the worst point x_{i_1}; (b) movement of the simplex in the two-dimensional space starting with $1 - 2 - 3$ points

new point is generated by reflection. A unit vector from x_{ik}^W through the centroid of the remaining points x_{i0} may be expressed as

$$u_i = \frac{x_{i0} - x_{ik}^W}{|x_{i0} - x_{ik}^W|}$$

so that the new point is given by

$$x_{ik}^{New} \text{ (reflected)} = \alpha |x_{i0} - x_{ik}^W| u_i + x_{i0} = (1 + \alpha) x_{i0} - x_{ik}^W.$$

x_{i0} is the centroid of starting points excluding the worst one, i.e.

$$x_{i0} = \frac{1}{m-1} \left(\sum_{k=1}^{m} x_{ik} - x_{ik}^W \right)$$

while $\alpha \geq 1$ is the reflection coefficient. If the reflection does not give a better point, contraction should be used:

$$x_{ik}^C = x_{i0} - \beta(x_{i0} - x_{ik}^R), \quad 0 < \beta < 1.$$

If x_{ik}^C (contracted) is not better than x_{ik}^R (reflected), we generate new points as follows:

$$x_{ik}^{New} = x_{ik}^{best} + x_{ik}^{old}/2.$$

If x_{ik}^R is the best point, we apply the expansion:

$$x_{ik}^E = x_{i0} + \gamma(x_{ik}^R - x_{i0}), \quad \gamma > 1.$$

If x_{ik}^E is not better than x_{ik}^R, we continue with x_{ik}^R.

The convergence criterion is

$$\left[\frac{1}{n} \left\{ \sum_{i=1}^n [f(x_{ik}) - f(x_{i0})]^2 - [f(x_{ik}^W) - f(x_{i0})]^2 \right\} \right]^{1/2} < \epsilon.$$

The movement of the simplex by sequential reflection of the worst points is illustrated in Fig. 2.8(b).

2.6 The Box Method

The Box Method is a modification of the simplex technique and is suitable for constrained problems with inequality constraints. It is not suitable for linear or equality constraints. The method is also known as the "complex" method, because it uses a set of $m \geq n + 1$ points, termed complex points.

For computational reasons it is advisable to use the same subscripts for the n explicit variables (x_i; $i = 1, \ldots, n$) and for the $p - n$ implicit variables (stress, deflection, etc., g_i; $i = n + 1, \ldots, p$). Thus, the explicit and implicit constraints can be written in the form

$$x_i^L \leq x_i \leq x_i^U \qquad i = 1, \ldots, n$$

$$g_i^L \leq g_i \leq g_i^U \qquad i = n + 1, \ldots, p.$$

The several steps of the algorithm are as follows (see flow chart in Fig. 2.9).

(1) For the first iteration cycle ($IT = 0$) the starting complex of $m \geq n + 1$ points is generated. This consists of a feasible starting point ($k = 1$) and $m - 1$ additional points generated from random numbers r_{ik} and the constraints for each of the independent variables:

$$x_{ik} = x_i^L + r_{ik}(x_i^U - x_i^L); \quad i = 1, \ldots, n.$$

r_{ik} are random numbers between 0 and 1. If, in subsequent calculations ($IT > 0$), the explicit constraints are violated, the point is moved a small distance δ inside

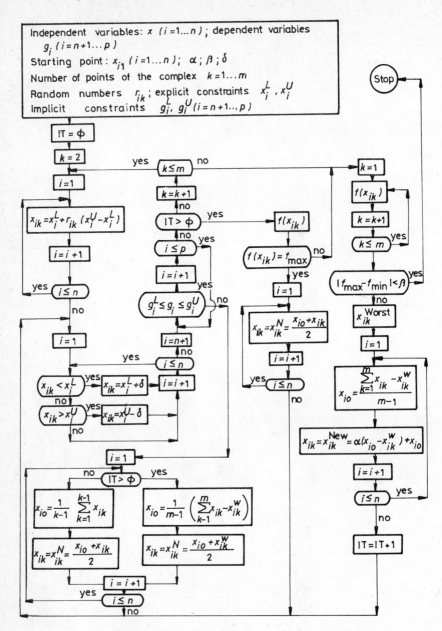

Independent variables: x $(i=1...n)$; dependent variables
g_i $(i=n+1...p)$
Starting point: x_{i_1} $(i=1...n)$; α; β; δ
Number of points of the complex $k=1...m$
Random numbers r_{ik}; explicit constraints x_i^L, x_i^U
Implicit constraints g_i^L, g_i^U $(i=n+1..,p)$

Fig. 2.9 – Flow chart of the Box method

the violated limit. If a point of the   complex fails to satisfy the implicit constraints, we calculate the centroid of the previous points

$$x_{i0} = \frac{1}{k-1} \sum_{k=1}^{k-1} x_{ik}$$

and then take a new point

$$x_{ik}^{New} = \frac{x_{i0} + x_{ik}^{old}}{2} .$$

All points of the complex should be checked ($k = 1, \ldots, m$).

(2) The objective function values are calculated and the convergence criterion checked: $|f_{max} - f_{min}| < \beta$. If the convergence criterion is not satisfied, the worst point x_{ik}^{Worst} (for which $f(x_{ik}^{W}) = f_{max}$) is rejected, it being replaced by a new one so that

$$x_{ik}^{New} = \alpha(x_{i0} - x_{ik}^{W}) + x_{i0}$$

where

$$x_{i0} = \frac{1}{m-1}\left(\sum_{k=1}^{m} x_{ik} - x_{ik}^{W} \right)$$

is the centroid of the remaining points. The proposed value for the reflexion coefficient is $\alpha = 1.3$.

(3) We now continue with the next iteration cycle ($IT = 1$). If the implicit constraints are not fulfilled, a new point is calculated:

$$x_{i0} = \frac{1}{m-1}\left(\sum_{k=1}^{m} x_{ik} - x_{ik}^{W} \right)$$

$$x_{ik}^{New} = \frac{x_{i0} + x_{ik}^{W}}{2} .$$

If all constraints are satisfied, but the point remains the worst one (i.e. $f(x_{ik}^{New}) = f_{max}$), a new point must be calculated:

$$x_{ik}^{New} = \frac{x_{i0} + x_{ik}^{W}}{2} .$$

If $f(x_{ik}^{New}) \neq f_{max}$, we continue with step 2 until the convergence criterion is fulfilled.

2.7 Backtrack Programming

The backtrack method solves nonlinear constrained function minimization problems by a systematic search procedure. This combinatorial discrete programming method can be successfully applied to optimization problems if the number of unknowns is not too large.

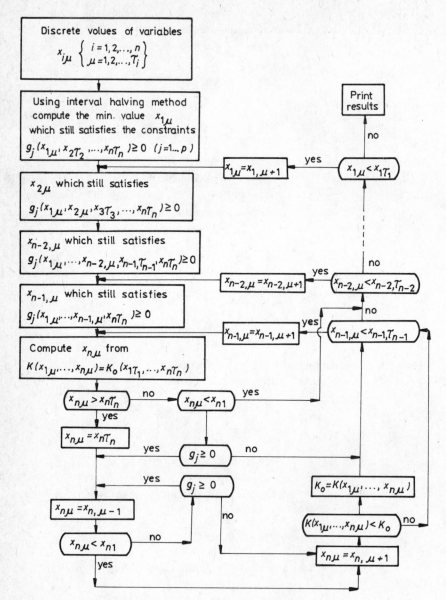

Fig. 2.10 – Flow chart of the backtrack method (cf. Table 2.1)

Table 2.1

Numerical Example Illustrating the Steps in the Backtrack Method in the Case
of Three Unknowns

Notations: + satisfied, − not satisfied

x_1 [cm]	x_2 [cm]	x_3 [cm^2]	K [cm^2]	g_1	g_2	Comments
74	0.9	22	110.6	+	+	$K_0 = 110.6$ cm^2
66	0.9	22	103.4	+	+	$x_{1\,min}$ is satisfactory, it is not necessary
66	0.5	22	77.0	+	−	to use the halving process for x_1, only
66	0.7	22	90.2	+	+	for x_2
66	0.6	22	83.6	+	+	
66	0.6	22	83.6	+	+	$x_3 = (110.6 - 66 \times 0.6)/2 = 35.5 > x_{3\,max}$;
66	0.6	21	81.6	+	+	$K_0 = 79.6$; backtrack with x_2
66	0.6	20	79.6	+	+	
66	0.6	19	77.6	−	+	
66	0.7	16	78.2	−	+	$x_3 = (79.6 - 66 \times 0.7)/2 = 16.7$
66	0.8	13				$x_3 < x_{3\,min}$; backtrack with x_1
68	0.9	22	105.2	+	+	Halving process for x_2
68	0.5	22	78.0	+	−	
68	0.7	22	91.6	+	+	
68	0.6	22	84.8	+	+	
68	0.6	19	78.8	+	+	$x_3 = (79.6 - 68 \times 0.6)/2 = 19.4$
68	0.6	18	76.8	−	+	$K_0 = 78.8$; backtrack with x_2
68	0.7	15	77.6	−	+	$x_3 = (78.8 - 68 \times 0.7)/2 = 15.5$
68	0.8	12				$x_3 < x_{3\,min}$; backtrack with x_1
70	0.9	22	107.0	+	+	Halving process for x_2
70	0.5	22	79.0	+	−	
70	0.7	22	93.0	+	+	
70	0.6	22	86.0	+	+	
70	0.6	18	78.0	+	+	$x_3 = (78.8 - 70 \times 0.6)/2 = 18.4$
70	0.6	17	76.0	−	+	$K_0 = 78.0$; backtrack with x_2
70	0.7	14	77.0	−	+	$x_3 = (78.0 - 70 \times 0.7)/2 = 14.5$
70	0.8	11				$x_3 < x_{3\,min}$; backtrack with x_1
72	0.9	22	108.8	+	+	Halving process for x_2
72	0.5	22	80.0	+	−	
72	0.7	22	94.4	+	+	
72	0.6	22	87.2	+	−	
72	0.7	13				$x_3 = (78.0 - 72 \times 0.7)/2 = 13.8 <$ $< x_{3\,min}$; backtrack with x_1
74	0.9	22	110.6	+	+	Halving process for x_2
74	0.5	22	81.0	+	−	
74	0.7	22	95.8	+	+	
74	0.6	22	88.4	+	−	
74	0.7	13				$x_3 = (78.0 - 74 \times 0.7)/2 = 13.1 < x_{3\,min}$ $x_1 = x_{1\,max}$; Results: $K_{0\,min} = 78.0$ $x_1 = 70, x_2 = 0.6; x_3 = 18$

The general exposition of backtrack can be found in the works of Walker (1960), Golomb and Baumert (1965) and Bitner and Reingold (1975). This method was applied to welded girder design by Lewis (1968) and Annamalai (1970). Farkas and Szabó (1980) have used it for the minimum cost design of hybrid I-beams. An estimation procedure for the efficiency of backtrack programming was proposed by Knuth (1975).

In the present work we describe only the special algorithm suitable for optimum design of welded structures which are characterized by monotonically increasing cost functions. Thus, the optimum solution can be found by decreasing the variables.

We should search for a vector of variables $\mathbf{x} = \left\{x_i\right\}$ $(i = 1, \ldots, n)$ for which the cost function $K(x_i)$ will be a minimum and which will also satisfy the design constraints $g_j(x_i) \geqslant 0$ $(j = 1, \ldots, p)$. For the variables, series of discrete values are given in increasing order as follows: $x_{i1}, \ldots, x_{i\mu}, \ldots, x_{i\tau_i}$ $(1 < \mu < \tau_i)$. In special cases the series may be determined by $x_{i\min}$, $x_{i\max}$ and by the constant steps Δx_i between them. The flow chart for the backtrack method is given in Fig. 2.10

Generally, a partial search is carried out for each variable and, if the possibilities are exhausted ("fathomed"), a backtrack is made and a new partial search is performed. The main phases of the calculation are as follows.

(1) With set constant values of $x_{i\tau_i}$ $(i = 2, \ldots, n)$ the minimum $x_{i\mu}$ value satisfying the design constraints is searched for. The search may be made more efficient by using the interval halving procedure. The latter method can be employed if the series of discrete values satisfies the following conditions:

$$x_{i\mu} - x_{i,\mu-1} = \Delta x_i$$

and

$$x_{i\tau_i} - x_{i1} = 2^q \Delta x_i$$

where q is an integer. The halving method is also applicable when the series of values used does not satisfy these conditions; however, the number of values is now $\tau_i = 2^q + 1$. In this more general case, a basic series $a_1, \ldots, a_k, \ldots, a_r$ $(r = 2^q + 1)$ can be defined (having constant steps $a_k - a_{k-1} = \Delta a$) and each value of the general series can then be adjoined to that of the basic series. Finally, in the case of a completely general series, the latter can be completed with values equal to $x_{i\tau_i}$ provided $\tau_i < r$. For instance, taking $\Delta a = 1$ and $q = 3$, the basic and the completed series are as follows:

Basic:	1,	2,	3,	4,	5,	6,	7,	8,	9
Completed:	4,	6,	8,	10,	12,	14,	16,	16,	16

The flow chart for the interval halving method in the case of a general series of discrete values is shown in Fig. 2.11.

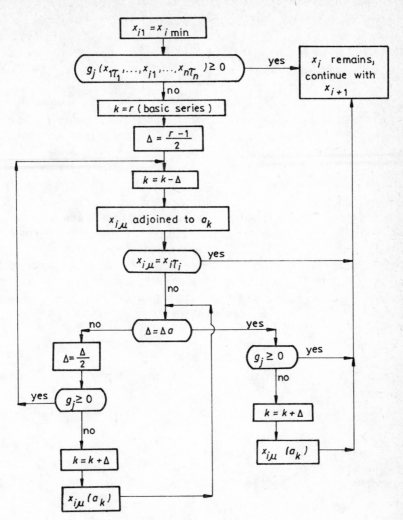

Fig. 2.11 – Flow chart of the interval halving method

(2) As in the case of the first phase, the halving process is now used with constant values $x_{1\mu}, x_{3\tau_3}, \ldots, x_{n\tau_n}$, and the minimum $x_{2\mu}$ value satisfying the design constraints is then determined.

(3) The least value $x_{n\mu}$ is calculated from the equation

$$K(x_{1\mu}, \ldots, x_{n\mu}) = K_0$$

where K_0 is the value of the cost function calculated by inserting the maximum x_i-values. Regarding the $x_{n\mu}$-value, three cases may occur as follows.

4

(3a) If $x_{n\mu} > x_{n\tau_n}$, we take $x_{n\mu} = x_{n\tau_n}$ and decrease it step-by-step till $x_{n\mu}$ which statisfies the constraints or till x_{n1} are reached. Then the partial search region is fathomed and we must backtrack to x_{n-1}. If $x_{n-1,\mu} < x_{n-1,\tau_{n-1}}$ we continue the calculation with $x_{n-1,\mu+1}$; if $x_{n-1,\mu} = x_{n-1,\tau_{n-1}}$, we backtrack to x_1.

(3b) If $x_{n\mu} < x_{n1}$, we backtrack to x_{n-2}.

(3c) If $x_{n1} < x_{n\mu} < x_{n\tau_n}$, and $x_{n\mu}$ does not satisfy the constraints, we backtrack to $x_{n-1,\mu}$. If the constraints are satisfied, we continue the calculation according to 3a.

The number of all possible variations is $\prod_{i=1}^{n} \tau_i$. However, the method investigates only a relatively small number of these. Since the efficiency of the method depends on many factors (number of unknowns, lists of discrete values, position of the optimum values in the lists, complexity of the cost function and/or that of the design constraints), it is difficult to predict the run time.

As an illustrative *example,* the minimum cross-sectional area design of a welded I-beam, subject to bending and compression, will now be considered (see also Section 6.3.3). This numerical example also shows how the various steps of the calculation can readily be performed with a simple pocket calculator.

The objective function is

$$K = A = ht_w + 2A_f$$

where $h = x_1$ (web height), $t_w = x_2$ (web thickness), $A_f = x_3$ (area of a flange). The stress constraint, g_1, is

$$\sigma_M + \sigma_N \leqslant R_u$$

where $\sigma_M = M/W_x$, $W_x \cong h(A_f + ht_w/6)$ being the section modulus, and $\sigma_N = N/A$, $N =$ compressive force. The constraint due to web buckling, g_2, may be expressed as

$$\frac{h}{t_w} \leqslant 145 \sqrt[4]{\frac{(1 + \sigma_N/\sigma_M)^2}{1 + 173(\sigma_N/\sigma_M)^2}}$$

in accordance to (A3.14) (see Appendix A3.2).

We take $M = 320$ kNm, $N = 128$ kN, $R_u = 200$ MPa. The lists of discrete values are then as follows (cm, cm^2):

	min	max	Δ
h	66	74	2
t_w	0.5	0.9	0.1
A_f	14	22	1

The total number of variations is $5 \times 5 \times 9 = 225$.

The steps of the calculations are shown in Table 2.1. The backtrack method requires only 36 tests for the obtention of the following optimum values: $h = 70$ cm, $t_w = 0.6$ cm, $A_f = 18$ cm^2.

It is worth noting that the results may depend on the sequence of variables, if these are linked to each other by some constraints. Some other considerations are described in Section 6.3.3.2 in connection with a numerical example with 8 variables.

The Fortran program of the backtrack technique is given in the Appendix. This program was applied (with some modifications) to other optimum design problems. Results of some of these are described in later chapters (Sections 3.2, 4.4, 5.2.8.4, 5.3.1, 5.4, 5.5.8, 6.3.3.2, 6.3.4.4).

PART II

APPLICATIONS OF OPTIMUM-DESIGN TECHNIQUES

Chapter 3

Compressed Struts
of Constant Square Hollow Section

3.1 Optimum Design Neglecting the Effects of Initial Imperfections, Residual Stresses and the Interaction of Overall and Local Buckling Modes

In the minimum cross-sectional area design of the two-pinned compressed strut shown in Fig. 3.1, the dimensions b and t should be optimized considering the constraints due to overall and local buckling. Shanley (1960) (as well as Spunt (1971), and Akita and Kitamura (1972)) have solved this problem (and the related problem of the optimum design of circular thin-walled tubes) without consideration of initial imperfections and residual welding stresses, and neglecting also the effect of interaction of the overall and local buckling modes. For purposes of comparison, we begin by discussing first this simplified approach.

(1) *Constraint due to overall buckling:* $N/(4bt) \leqslant \sigma_{cr\,o}$.

In the elastic region

$$\sigma_{cr\,o} = \pi^2 E/\lambda^2 \quad \text{for} \quad \sigma_{cr\,o} \leqslant R_y/2. \tag{3.1}$$

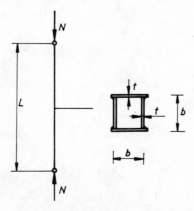

Fig. 3.1 – Compressed strut of welded square box cross-section

In the plastic region

$$\sigma_{cr\,o} = R_y - \frac{R_y^2}{4E\pi^2/\lambda^2} \qquad \text{for} \qquad \sigma_{cr\,o} > R_y/2 . \tag{3.2}$$

(2) *Constraint due to local buckling:* $N/(4bt) \leqslant \sigma_{cr\,1}$.

In the elastic region

$$\sigma_{cr\,1} = \frac{\pi^2 E}{3(1 - \nu^2)} \left(\frac{t}{b}\right)^2 \qquad \text{for} \qquad \sigma_{cr\,1} \leqslant R_y/2 . \tag{3.3}$$

In the plastic region

$$\sigma_{cr\,1} = R_y - \frac{3(1 - \nu^2)R_y^2}{4E\pi^2} \left(\frac{b}{t}\right)^2 \qquad \text{for} \qquad \sigma_{cr\,1} > R_y/2 . \tag{3.4}$$

For the case of a square tube the slenderness is given by

$$\lambda = L/i = \sqrt{6}\,L/b. \tag{3.5}$$

Instead of b and t we introduce the following dimensionless parameters:

$$x_1 = \overline{\lambda}^2 = \frac{6R_y}{\pi^2 E} \left(\frac{L}{b}\right)^2 = p^{-1}\left(\frac{L}{b}\right)^2 ; \qquad p = \frac{\pi^2 E}{6R_y} \tag{3.6}$$

and

$$x_2 = \frac{b}{t}\,q^{-1} = \vartheta q^{-1} ; \qquad q = \sqrt{\frac{\pi^2 E}{3(1 - \nu^2)R_u}} . \tag{3.7}$$

With the notation

$$c_N = \frac{N}{4L^2 R_u}\,pq \tag{3.8}$$

the constraints $(3.1) - (3.4)$ can be written in the following form:

$$\frac{1}{x_1} - c_N x_1 x_2 \geqslant 0 \qquad \text{for} \qquad x_1 \geqslant 2 \tag{3.9}$$

$$1 - \frac{x_1}{4} - c_N x_1 x_2 \geqslant 0 \qquad \text{for} \qquad x_1 < 2 \tag{3.10}$$

$$\frac{R_u}{R_y x_2^2} - c_N x_1 x_2 \geqslant 0 \qquad \text{for} \qquad x_2 \geqslant \sqrt{2R_u/R_y} \tag{3.11}$$

$$1 - \frac{R_y}{4R_u}\,x_2^2 - c_N x_1 x_2 \geqslant 0 \qquad \text{for} \qquad x_2 < \sqrt{2R_u/R_y} . \tag{3.12}$$

Equating (3.9) to (3.11), and (3.10) to (3.12), we obtain

$$x_{1\,opt} = \frac{R_y}{R_u}\,x_{2\,opt}^2 . \tag{3.13}$$

Assuming that the case being investigated lies in the elastic region, we get (on the basis of (3.11) and (3.13))

$$x'_2 \text{ opt} = \sqrt[5]{\frac{R_u^2}{c_N R_y^2}} . \tag{3.14}$$

If $x'_2 \text{ opt} < \sqrt{2R_u/R_y}$, the solution (3.14) is not valid because the case lies in the plastic zone and the optimum may then be found by solving the following equation derived by means of (3.12) and (3.13):

$$1 - \frac{R_y}{4R_u} x_2^2 - \frac{c_N R_y}{R_u} x_2^3 = 0. \tag{3.15}$$

(3) *Numerical example*

$$N/L^2 = 10^{-2} \text{ MPa} ; \quad L = 5 \text{ m} ; \quad E = 210 \text{ GPa} ;$$

$$R_u = 190 \text{ MPa} ; \quad R_y = 240 \text{ MPa} .$$

$$p = 1429.3 ; \quad q = 63.2123 ; \quad c_N = 1.1971 .$$

With (3.14) we get $x'_2 \text{ opt} = 0.8786 < 1.2583$; thus we must solve Eq. (3.15), i.e.

$$x_2^2 (0.31579 + 1.1971 x_2) = 1.$$

The result is $x_2 \text{ opt} = 0.8616$. Finally, the following values are obtained: $x_1 \text{ opt} = 0.93774$; $L/b = x_1 \text{ opt} p = 36.74$; $b = 136.1$ mm; $b/t = q x_2 \text{ opt} = 54.46$; $t = 2.5$ mm; $A = 1361$ mm^2; $\sigma_N = 250\ 000/1361 = 183.7$ MPa; $10^4 A/L^2 = 0.544$.

3.2 Optimum Design Considering the Effects of Initial Imperfections and Residual Welding Stresses, but Neglecting the Interaction of Buckling Modes

The author has proposed a relatively simple optimum design procedure (Farkas 1977c). By using (3.6) and (3.7) the objective function is

$$A = 4 bt = \frac{4L^2}{pqx_1 x_2} . \tag{3.16}$$

For the constraint imposed by overall buckling the European buckling curves are applied (see Appendix A2). By using (A2.1) and (A2.2), this constraint may be written as

$$\frac{Npq}{4L^2} x_1 x_2 \leqslant \frac{R_u}{0.5 + \alpha_1 x_1 + \sqrt{(0.5 + \alpha_1 x_1)^2 - \beta_1 x_1}}. \tag{3.17}$$

In the *constraint due to local buckling*

$$N/A \leqslant \psi R_u$$

the modified Faulkner's effective width formula is used (see Appendix A3). Thus, both criteria take into account the effect of initial imperfections and residual welding stresses, although the interaction of overall and local buckling is neglected.

Using the definition of (A3.4), with $\nu = 0.3$, we get

$$\lambda_p = x_2 q\sqrt{R_u/E} = 1.9 x_2 . \tag{3.18}$$

The constraint valid for the elastic region can then be written as (see Eq. (A3.6))

$$\frac{Npq}{4L^2 R_u} x_1 x_2 \leqslant \frac{2}{1.9 x_2} - \frac{1}{1.9^2 x_2^2} - \frac{\sigma_c(\vartheta)}{R_{y1}} \quad \text{for} \quad x_2 \geqslant x_{20} = \sqrt{R_u/R_e}$$

$$\tag{3.19}$$

while that for the plastic region (Eq. (A3.8)) is

$$\frac{Npq}{4L^2 R_u} x_1 x_2 \leqslant 1 - (1 - \psi_0)\left(\frac{x_2}{x_{20}}\right)^2 \quad \text{for} \quad 0 \leqslant x_2 \leqslant x_{20} . \tag{3.20}$$

ψ_0 can be calculated by (A3.8). By using (A1.11) we get

$$\frac{\sigma_c}{R_{y1}} = \frac{A_T t}{\vartheta \epsilon_{y1} t^2} . \tag{3.21}$$

Furthermore, on the basis of (A1.3) and (A1.13), for fillet welds of dimension $a_w = 0.7t$ and for $R_{y1} = 300$ MPa ($\epsilon_{y1} = 1.5 \times 10^3$), one obtains

$$\sigma_c/R_{y1} = 16.3464/\vartheta . \tag{3.22}$$

The solution of this two-dimensional problem can be obtained graphonumerically. Computations were carried out by using both the SUMT and the backtrack algorithms.

(1) *Numerical example solved graphonumerically*

The data are the same as those in Section 3.1. From Table A2.1, $\alpha_1 = 0.65$ and $\beta_1 = 0.80$. According to (3.19), $x_{20} = 1.14867$. From (A3.5) and taking $\sigma_{max} = R_u$, $\lambda_{p0} = 2.1825$. Using (A3.8), $C_1/\vartheta_0 = 0.22528$ and $\psi_0 = 0.48116$. Eqs (3.17), (3.19) and (3.20) then become

$$x_2 \leqslant \frac{0.83533L^2/N}{x_1[0.5 + 0.65x_1 + \sqrt{(0.5 + 0.65x_1)^2 - 0.8x_1}]} \qquad (3.23)$$

$$x_1 \leqslant \frac{0.66265L^2/N}{x_2^2}\left(1 - \frac{0.34870}{x_2}\right) \quad \text{for} \quad x_2 \geqslant 1.14867 \quad (3.24)$$

$$x_1 \leqslant \frac{1 - 0.39339x_2^2}{1.19712x_2}\left(\frac{L^2}{N}\right) \qquad \text{for} \quad x_2 \leqslant 1.14867. \quad (3.25)$$

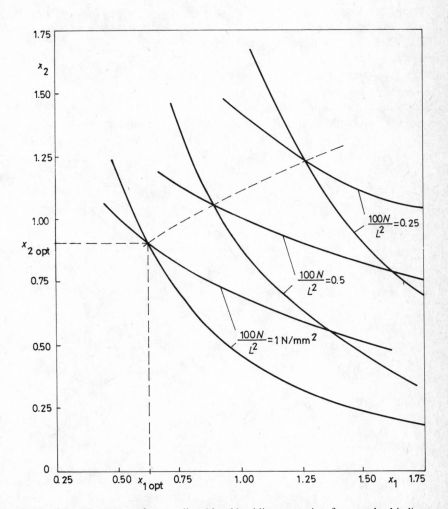

Fig. 3.2 – Limit curves for overall and local buckling constraints for some load-indices; the dotted line contains the optimum points

Figure 3.2 shows the limit curves for the constraints due to overall and local buckling in the $x_1 - x_2$ system, for load-index values of $100N/L^2 = 1; 0.5; 0.25$ MPa. The points of optima can be obtained by intersecting the corresponding lines of constraints with a given value of N/L^2. The dotted line contains the intersection points. It may be shown numerically that, at the intersection points, the Kuhn–Tucker conditions are satisfied so that the A_{min}-criterion is also fulfilled.

For the data $N/L^2 = 10^{-2}$ MPa and $L = 5$ m (used in the numerical example of Section 3.1) the graphical solution gives $x_{1\,opt} = 0.632$, $x_{2\,opt} = 0.900$. The remaining values are as follows: $b = 165.8$ mm; $\vartheta = 56.89$; $t = 2.91$ mm; $\sigma_N = 129.5$ MPa; $10^4\,A/L^2 = 0.772$.

It can be seen that, in this numerical example, the neglecting of the effects of initial imperfections and residual welding stresses results in a cross-sectional area 30% smaller than that obtained by the author's method.

(2) *Optimum cross-sections determined by the SUMT method*

By using the Fortran program for SUMT (see Section 2.3) we have obtained optimum cross-sections for nine values of N/L^2. The results appear in Fig. 3.3.

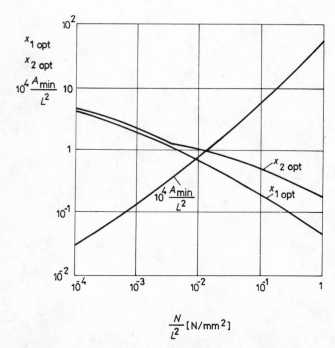

$$\frac{N}{L^2}\,[\text{N/mm}^2]$$

Fig. 3.3 – Characteristics of optimum sections obtained by the SUMT method

(3) *Optimum cross-sections obtained by the backtrack method*

By using the Fortran program developed for the backtrack method (see Section 2.7) a number of optimum sections were computed. The constraint imposed by the overall buckling criterion was as follows (see Appendix A2.7):

$$N/A \leqslant \varphi_b R_u$$

$$\varphi_b = \frac{1}{2\bar{\lambda}^2}\left[1 + \eta_b + \bar{\lambda}^2 - \sqrt{(1 + \eta_b + \bar{\lambda}^2)^2 - 4\bar{\lambda}^2}\right]$$

$$\eta_b = 0.281\sqrt{\bar{\lambda}^2 - 0.04}.$$

The series of discrete values were taken as follows:

	min	max	Δ
N [kN]	100	725	25
L [m]	2.4	4.6	0.2
b [mm]	40	600	10
t [mm]	1	35	1

Results are given in Appendix A7. These suboptimized rods can be used in the optimum design of trusses (see Section 4.4).

3.3 Interaction of Overall and Local Buckling Modes

A number of articles have shown that compressed struts optimized by neglecting the effects of initial imperfections and residual stresses (refer to Section 3.1) are in fact very sensitive to these effects. Furthermore, it is well known that local buckling can interact detrimentally with the column buckling mode. The analysis of this interaction was treated, for example, by Škaloud (1962), Skaloud and Zörnerová (1970), Škaloud and Náprstek (1977). The paper of Svensson and Croll (1975) tackles this problem using the van der Neut's model, but neglecting the effect of residual stresses. Klöppel and Schubert (1971), on the other hand, have solved the problem by allowing for both initial imperfections and residual welding stresses. Little (1979) has published a detailed analysis as well as tables of optimized dimensions of welded box sections, while Braham *et al.*(1980) have proposed a calculation method for cold-formed rectangular hollow sections (see Appendix A2).

We have computed a series of optimum square hollow sections by backtrack programming on the basis of the formulae by Braham *et al.* (1980). The objective function is expressed as

$$A = 4t\left[b - r\left(2 - \frac{\pi}{2}\right)\right]$$

and the constraint is written in the form

$$\frac{N}{A} \leqslant R_y \, \overline{N}' \, \overline{N}_v.$$

$\overline{N}'$ and $\overline{N}_v$ may be calculated with formulae (A2.8) − (A2.12), given in Appendix A2. The following discrete values have been considered (in mm):

B: 40, 60, 80, 100, 120, 140, 160, 180, 200, 220, 250, 285, 300, 325, 350, 400

t: 1.2, 1.6, 2.0, 2.6, 3.2, 4.0, 5.0, 6.3, 7.1, 8.0, 10.0, 12.5

Table 3.1

Optimum Dimensions of Compressed Columns of Cold-Finished Square Hollow Section Computed by the Method of Braham *et al.* (1980). $R_y = 355$ MPa, $\alpha_B = 0.489$, $\alpha_v = 0.67$

N [kN]	L [m]	B × t [mm]	N [kN]	L [m]	B × t [mm]	N [kN]	L [m]	B × t [mm]	N [kN]	L [m]	B × t [mm]
	2	60 × 1.2		2	80 × 3.2		2	140 × 4.0		2	220 × 7.1
	4	100 × 1.2		4	140 × 2.6		4	160 × 5.0		4	220 × 8.0
30	6	120 × 1.2	200	6	180 × 2.6	600	6	200 × 5.0	1800	6	285 × 7.1
	8	140 × 1.2		8	180 × 3.2		8	200 × 6.3		8	300 × 8.0
	10	160 × 1.2		10	200 × 3.2		10	285 × 5.0		10	350 × 8.0
	2	60 × 1.6		2	100 × 2.6		2	160 × 4.0		2	180 × 10.0
	4	100 × 1.2		4	140 × 3.2		4	180 × 5.0		4	250 × 8.0
40	6	140 × 1.2	250	6	140 × 4.0	700	6	200 × 5.0	2000	6	285 × 8.0
	8	160 × 1.2		8	200 × 3.2		8	250 × 5.0		8	325 × 8.0
	10	200 × 1.2		10	200 × 4.0		10	300 × 5.0		10	325 × 10.0
	2	60 × 1.6		2	100 × 3.2		2	160 × 5.0		2	220 × 8.0
	4	100 × 1.6		4	140 × 3.2		4	180 × 5.0		4	250 × 8.0
60	6	140 × 1.6	300	6	160 × 4.0	800	6	200 × 6.3	2200	6	285 × 8.0
	8	160 × 1.6		8	200 × 4.0		8	220 × 6.3		8	350 × 8.0
	10	180 × 1.6		10	220 × 4.0		10	250 × 7.1		10	325 × 10.0
	2	80 × 1.6		2	120 × 3.2		2	180 × 5.0		2	250 × 8.0
	4	100 × 2.0		4	140 × 4.0		4	180 × 6.3		4	285 × 8.0
80	6	140 × 2.0	350	6	180 × 4.0	1000	6	220 × 6.3	2500	6	285 × 10.0
	8	160 × 2.0		8	200 × 4.0		8	250 × 6.3		8	325 × 10.0
	10	180 × 2.0		10	200 × 5.0		10	300 × 6.3		10	350 × 10.0
	2	80 × 2.0		2	120 × 3.2		2	140 × 8.0		2	285 × 8.0
	4	120 × 2.0		4	140 × 4.0		4	200 × 6.3		4	300 × 8.0
100	6	120 × 2.6	400	6	180 × 4.0	1200	6	220 × 7.1	2800	6	300 × 10.0
	8	180 × 2.0		8	220 × 4.0		8	250 × 7.1		8	325 × 10.0
	10	200 × 2.0		10	250 × 4.0		10	300 × 7.1		10	400 × 10.0
	2	80 × 2.6		2	120 × 4.0		2	180 × 7.1		2	285 × 8.0
	4	120 × 2.6		4	160 × 4.0		4	220 × 7.1		4	285 × 10.0
150	6	160 × 2.6	500	6	220 × 4.0	1500	6	250 × 7.1	3000	6	325 × 10.0
	8	180 × 2.6		8	220 × 5.0		8	285 × 7.1		8	350 × 10.0
	10	180 × 3.2		10	250 × 5.0		10	300 × 8.0		10	400 × 10.0

The results are summarized in Table 3.1. It should be noted that, according to the efficiency chart proposed by Usami and Fukumoto (1982), for all optimum sections $\bar{\lambda}_v \geqslant 0.8$. For some sections in Table 3.1 $\bar{\lambda}_v < 0.8$ because of the coarse series of discrete values.

The same problem was solved by using author's method explained in Section 3.2. The comparison of the numerical results obtained by the two methods in the case of $R_y = 235$ MPa shows that the author's method gives suitable values

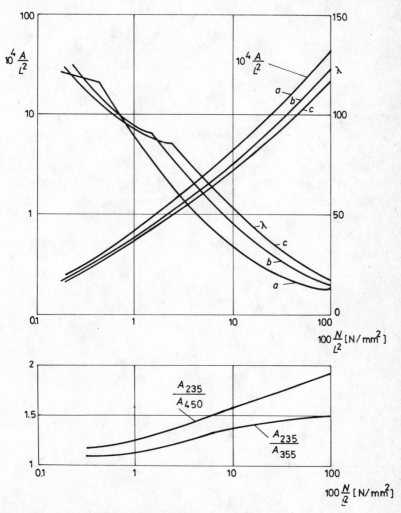

Fig. 3.4 – A_{min}/L^2 and λ values for welded square box cross-sections in the case of steels of yield stress $R_y = 235$ (a), 355 (b) and 450 MPa (c), respectively. The A_{235}/A_{355} and A_{235}/A_{450} ratios characterize the weight reduction resulting from the increasing of the yield stress

for the force/length ratios of $100\,N/L^2 \geqslant 5$ MPa. In the case of lesser $100\,N/L^2$ values, the difference between the optimum cross-sectional areas computed by the two methods is about $10 - 15\%$.

The diagrams of A/L^2 versus N/L^2 can be obtained by hand calculations using the efficiency chart proposed by Usami and Fukumoto (1982). In the case of a welded square box cross-section, using the formulae

$$A = 4bt; \quad i = b/\sqrt{6}; \quad \lambda = L/i \quad \text{and} \quad \overline{\lambda}_\nu = \frac{b}{1.9t}\sqrt{\frac{R_y}{E}}\,,$$

$\overline{\lambda}'$ can be expressed with A, L and $\overline{\lambda}_\nu$:

$$\overline{\lambda}'^2 = \frac{\lambda^2 R_y \overline{N}_\nu}{\pi^2 E} = 1.27984 \left(\frac{R_y}{E}\right)^{3/2} \frac{L^2 \overline{N}_\nu}{A\,\overline{\lambda}_\nu}\,.$$

For given values of A, L, R_y and E, calculating $\overline{N}'$ and $\overline{N}_\nu$ with formulae (A2.8) and (A2.12), the efficiency $\overline{N}'\overline{N}_\nu$ can be plotted as a function of $\overline{\lambda}_\nu$. Then, determining $(\overline{N}'\overline{N}_\nu)_{max}$, the parameter $N/L^2 = R_y(\overline{N}'\overline{N}_\nu)_{max}\,(A/L^2)$ can be calculated. Figure 3.4 shows the diagrams for $R_y = 235,355$ and 450 MPa, calculated with $\alpha_\nu = 0.67$ and $\alpha_B = 0.335, 0.206$ and 0.125, respectively.

The A_{235}/A_{355} and A_{235}/A_{450} curves characterize the economy of the use of steels of higher yield stress. It can be seen that for larger λ values the economy decreases considerably.

Chapter 4

Trusses

4.1 Aspects of Structural Synthesis

Trusses may be *planar* or *spatial*. Some spatial trusses may be partitioned to planar components, others form the so-called lattice structures in the shapes of plates or shells. Trusses may be statically determinate or indeterminate relating to external reactions and/or internal forces.

Bars can be constructed from rolled, cold-formed or welded (open or closed) profiles. Circular or rectangular hollow section components are often welded together without using gusset plates. In the design of bars only stresses due to tensile or compressive internal forces are usually considered, although bending (or "secondary") stresses, introduced by the rigidity of joints, can be significant. In the design of chord and branch members the *failure modes of joints* (tearing of welds, punching shear of the chord face, shear failure of the chord sidewalls, local buckling of plate elements) should be taken into account (see, e.g. Wardenier 1982), Design Recommendations 1982)).

Compression members should be designed against overall buckling. In order to simplify the optimum design procedure, some authors prescribe an allowable stress value for compression members. This concept may only be used in approximate calculations, since the members usually lie in the plastic buckling zone. In general, it is more advisable to use suboptimized series of compressed struts (see Chapter 3).

The mass or cost of the structure may be taken as the objective function. The fabrication costs of various types of joints vary widely. Furthermore, the fabrication of joints is expensive, and thus it is advisable to use standardized joints. It is also clear that it would be uneconomical to fabricate all members from different sizes; thus, on the basis of preliminary, approximate calculations, groups of members of the same size can be determined. When designing the chords, one should bear in mind that these should be constructed from elements longer than the distances between joints.

Design constraints are usually as follows: stress and size criteria for each member (prescription of maximum stress, minimum size or maximum slenderness ratio), displacement criteria for one or more joints, and natural frequency criteria for the whole truss.

Variables are usually the cross-sectional areas of the members. In shape optimization the coordinates of the joints are also taken as variables. When the trusses have parallel chords the height (distance between the chords) may be optimized.

The optimization of joint positions may be illustrated by the simple example of the two-bar structure shown in Fig. 4.1.

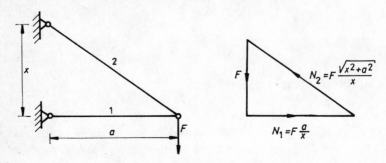

Fig. 4.1 – A simple example of the optimization of geometry

The volume to be minimized is

$$V = A_1 a + A_2 \sqrt{x^2 + a^2}.$$

Taking into account only stress constraints with constant limit stresses R_{u1} and R_{u2} respectively, the cross-sectional areas may be expressed as

$$A_1 = Fa/(xR_{u1}), \qquad A_2 = F\sqrt{x^2 + a^2}/(xR_{u2}).$$

Inserting these expressions into the formula for the volume one obtains

$$V = \frac{Fa^2}{xR_{u1}} + \frac{F}{R_{u2}}\left(x + \frac{a^2}{x}\right).$$

The condition $dV/dx = 0$ then gives

$$x_{opt} = a\sqrt{1 + R_{u2}/R_{u1}}.$$

4.2 Literature Survey

A brief survey of some of the publications relating to the optimum design of trusses is in this section given. Some of the data are summarized in Table 4.1 while Figs 4.2 and 4.3 show the types of trusses usually investigated.

The statically indeterminate three-bar structure (Fig. 4.2a) is treated in many studies. Schmit (1960) has investigated the structural synthesis for this structure as a nonlinear mathematical programming problem. De Silva and Grant (1973)

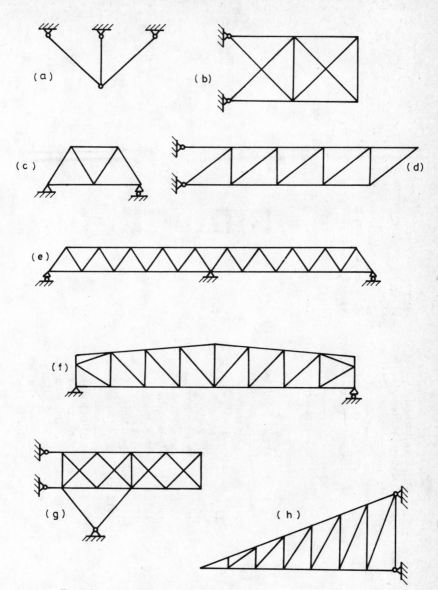

Fig. 4.2 – Some planar trusses studied in the literature (see Table 4.1)

have compared some nonlinear programming methods using the results relating to the same example. A number of authors have treated the 10-bar planar truss, as well as the 4-, 25- and 72-bar spatial trusses shown in Figs 4.2, 4.3 mainly for purposes of comparing the various optimization techniques.

Table 4.1 contains relevant information regarding the mathematical method used, showing also whether the study optimizes the truss geometry, and whether consideration is given to the constraints of overall buckling and/or frequency.

5*

Table 4.1

Data for Some of the Trusses Studied in the Literature

Abbreviations: LP – linear programming; SLP – sequential LP; sep. LP – separable LP; GP – geometric programming; DP – dynamic programming; DFP – see Sect. 2.2; seq. DFP – sequential DFP; Box – see Sect. 2.6; SUMT – see Sect. 2.5; OC – optimality criteria; grad. – gradient method; anal. – analytical method; const. – constant; var. – variable; var. h. – variable height; stab. – with stability constraints; fr. – with frequency constraints; comb. – combinatorial; subopt. – with suboptimized sections; di. – discrete; graph. – graphical; whole ind. build. – whole industrial building structure; combined – combined method; el + pl – elastic and plastic analysis

Reference		Truss type investigated	Geometry	Applied math. method	Comment
Schmit	1960	Fig. 4.2(a)	const.	grad.	stab.
Fox and Schmit	1966	11-bar planar, Figs 4.3(d), (e)	const.	seq. DFP	stab.
Rosenkranz	1968	Planar with parallel chords	var. h.	anal.	
Toakley	1968	15-bar planar	const.	more m.	di. var.
Felton and Hofmeister	1968	Fig. 4.3(d)	const.	grad.	subopt.
Icerman	1969	12-, 14-bar planar	const.	OC	fr.
Fox and Kapoor	1970	5-, 7-bar, Fig. 4.2	const.	grad.	fr.
Ladyzhenskiy	1970	Planar with parallel chords	var. h.	anal.	
Venkayya	1971	Figs 4.2(a), (b); 4.3(c), (e), (f)	const.	OC	
Lipson and Russell	1971	Whole ind. build.	var.	Box	stab.
LaPay and Goble	1971	Figs 4.2(a), (b)	const.	SUMT	di. var.
Cella and Logcher	1971	Fig. 4.2(g)	const.	comb.	
Majid and Anderson	1972	Fig. 4.2(b)	const.	sep. LP	
Vanderplaats and Moses	1972	Figs 4.3(a) – (b), (e)	var.	grad.	stab.
Morris	1972	6-bar planar	const.	GP	
Pickett *et al.*	1973	6-, 12-bar planar; Figs 4.3(e) –(f); 200-bar	const.	SUMT + DFP	
Pedersen	1973	Spherical cupola	var.	SLP	stab.
Koch	1973	Fig. 4.2(e), crane	var. h.	grad.	stab.
Lipson and Agrawal	1974	Figs 4.2(a), 4.3(a) – (b)	var.	Box	stab.
Majid	1974	More planar, Fig. 4.3(c)	var.	graph.	
Farkas	1974a	With parallel chords	var.	anal.	stab.

Author	Year	Description	Method	Type	Notes
Funaro	1974	Figs 4.2(c), (d)	sep. LP	const.	
Shaw	1974	Fig. 4.2(h)	OC	const.	fr.
Agha and Nelson	1976	Figs 4.2(b), 4.3(c), (e) (f)	SLP	const.	
Schmit and Miura	1976	Figs 4.2(b), 4.3(e), (f)	NEWSUMT	const.	
Lipp	1976	Figs 4.2(b), 4.3(f), 200-bar	OC	const.	
Dobbs and Nelson	1976	Figs 4.2(b), 4.3(e)	DP	var.	
Reitman and Shapiro	1976	With parallel chords	OC	const.	
Rizzi	1976	Figs 4.2(b), 4.3(e)	SUMT + DFP	var.	stab.
Thomas and Brown	1977	Fig. 4.2(b), whole ind. build	Box	var.	fr.
Lipson and Gwinn	1977	Fig. 4.3(e)	grad.	const.	
Feng et al.	1977	7-bar planar	DP	var.	stab.
Mitra et al.	1978	Transmission towers	NEWSUMT	const.	stab.
Schmit and Ramanathan	1978	50-bar planar	OC	const.	stab.
Ol'kov and Antipin	1978	Fig. 4.3(c)	OC	const.	stab.
Isreb	1978	Figs 4.2(a), 4.3(e)	OC	const.	
Khan et al.	1979	Figs 4.2(b), 4.3(c), (e), (f)	OC	const.	
Fleury	1979	Fig. 4.2(b)	OC	const.	combined
Schmit and Fleury	1980	Figs 4.2(b), 4.3(e), (f)	dual	const.	ACCESS 3
Harless	1980	Figs 4.2(a), (b); 4.3(c), (e)	spec. grad.	const.	
Kirsch and Benardout	1980	Figs 4.2(a), (b)	SLP	const.	
Imai and Shoji	1981	Fig. 4.2(b)	dual	const.	
Imai and Schmit	1981	18-bar, 23-bar	multiplier	var.	
Khot	1981	200-bar	OC	const.	
Svanberg	1981	3-bar pyramid, 39-bar tower	OC dual	var.	stab.
Grierson and Schmit	1982	Fig. 4.3(c), 22-bar spatial	OC	const.	el. and pl.

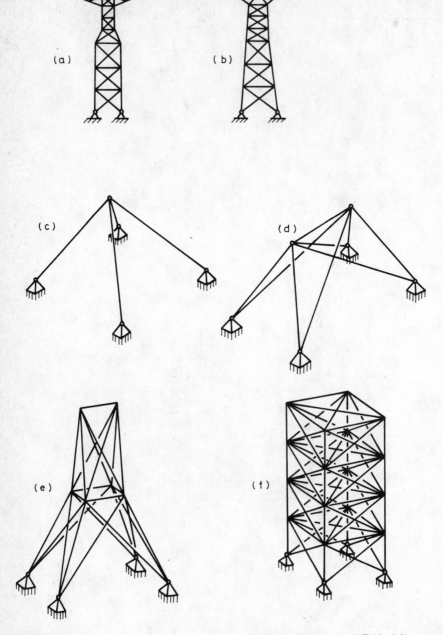

Fig. 4.3 – Some planar and spatial trusses studied in the literature (see Table 4.1)

Lipson and Russell (1971), and Thomas and Brown (1977) have studied the optimization of a whole industrial building structure. The books of Majid (1974) and Reitman and Shapiro (1976) contain some simple examples of truss optimization. Felton and Hofmeister (1968) treat the optimum design problem of a 9-bar spatial truss considering suboptimized members of circular hollow cross-section.

4.3 Application of Optimality-Criteria Methods

In appyling the optimality-criteria methods considered in Section 2.4 different solution algorithms may be used depending on the actual type of constraint(s).

4.3.1 Prescription of a Single Displacement Constraint

The material cost function of a rod structure is given by

$$K(A_i) = \sum_{i=1}^{n} k_i l_i A_1 \tag{4.1}$$

where k_i is the cost factor, l_i the bar length, and A_i, the cross-sectional area. The constraints are written in the form

$$g_j(A_i) \leq 0 \qquad\qquad j = 1, \ldots, p. \tag{4.2}$$

According to (2.26), the optimality criteria are as follows:

$$\left.\begin{array}{c} k_i l_i + \displaystyle\sum_{j=1}^{p} \lambda_j \frac{\partial g_j}{\partial A_i} = 0 \\[4mm] \lambda_j \geq 0 ; \quad \lambda_j g_j = 0 . \end{array}\right\} \tag{4.3}$$

The displacement constraints may be written as

$$g_j(A_i) = \sum_{i=1}^{n} \frac{e_{ij}}{A_i} - g_j^* \tag{4.4}$$

where g_j^* are the limit (allowable) values, while e_{ij} is given by

$$e_{ij} = \mathbf{f}_i^T \, \mathbf{r}_i \, \mathbf{q}_i^j . \tag{4.5}$$

e_{ij} represents the virtual strain energy in the ith member due to the jth virtual load vector $\mathbf{Q}^j$ and the applied load vector $\mathbf{F}$, while $\mathbf{f}$ and $\mathbf{q}$ are the displacement vectors corresponding to $\mathbf{F}$ and $\mathbf{Q}$ respectively. Finally, $\mathbf{r}_i$ is the stiffness matrix.

For bar structures

$$e_{ij} = \frac{N_i \, n_i^j \, l_i}{E_i}.$$ (4.6)

N_i and n_i^j are the internal forces in the ith member due to the applied load and the virtual load for the jth constraint respectively. While in statically determinate structures e_{ij} is constant, in indeterminate ones it depends on A_i, although it may also be taken as a constant for such structures as well.

Thus, for displacements constraints, the optimality criteria can be written as

$$\sum_{j=1}^{p} \lambda_j \, \frac{e_{ij}}{k_i \, l_i \, A_i^2} = 1$$

$$\lambda_j \geqslant 0 \, ; \qquad \lambda_j g_j = 0 \, .$$ (4.7)

For a single displacement constraint ($j = 1$) of

$$\sum_{1}^{n} e_i / A_i = w^*$$ (4.8)

the optimality criterion is given by

$$\lambda \, \frac{e_i}{k_i \, l_i \, A_i^2} = 1$$ (4.9)

so that

$$A_i = \sqrt{\lambda} \, \sqrt{\frac{e_i}{k_i \, l_i}} \, .$$ (4.10)

Inserting (4.10) into (4.8) one obtains

$$\sum_{1}^{n} \frac{\sqrt{e_i \, k_i \, l_i}}{\sqrt{\lambda}} = w^*$$ (4.11)

which gives

$$\sqrt{\lambda} = \frac{1}{w^*} \sum_{1}^{n} e_i \, k_i \, l_i \, .$$ (4.12)

Finally, inserting (4.12) into (4.10) yields

$$A_k = \frac{1}{w^*} \sqrt{\frac{e_k}{k_k \, l_k}} \sum_{1}^{n} \sqrt{e_i \, k_i \, l_i} \, .$$ (4.13)

For statically determinate trusses with members made of a common material:

$$e_i = N_i \, n_i \, l_i / E \, .$$ (4.14)

Hence (4.13) becomes

$$A_k = \frac{\sqrt{N_k \, n_k}}{w^* E} \sum_{l}^{n} l_i \sqrt{N_i \, n_i} \; . \tag{4.15}$$

The members can be *active* $(i \in m_a)$ or *passive* $(i \in m_p)$. With reference to displacement constraints an element is *passive* if

(1) A_k obtained from (4.15) is less than that determined through other constraints such as stress or size;
(2) $N_k \, n_k < 0$;
(3)   the kth element belongs to a group with the same size A_{max} and $A_k < < A_{max}$.

Since it is not known in advance whether an element is active or passive, an iteration should be applied. Considering passive members as well, (4.8) is modified as follows:

$$\sum_{m_a} \frac{e_i}{A_i} = w^* - w_{pass} \; ; \quad w_{pass} = \sum_{m_p} \frac{N_i \, n_i \, l_i}{A_i E} \; . \tag{4.16}$$

Thus (4.15) becomes

$$A_k = \frac{\sqrt{N_k \, n_k}}{E(w^* - w_{pass})} \sum_{m_a} l_i \sqrt{N_i \, n_i} \; . \tag{4.17}$$

4.3.2 Stress Constraints

Stress constraints may be written in the form

$$|N_i| / A_i \leqslant R_{ui} \tag{4.18}$$

where R_{ui} are the limiting (allowable) stresses for compressed bars which depend on the slenderness ratio.

It is evident that, in the case of statically determinate structures, all the constraints of (4.18) must be treated as active if the minimum cost defined by (4.1) is to be attained; i.e. the so-called *fully stressed design* (FSD) gives the minimum cost. Although the FSD does not correspond to the minimum cost of a statically indeterminate structure with varying load conditions, it can still be regarded as a good approximation to the minimum cost design. It was shown by Barta (1957) that each subsidiary statically determinate form of an indeterminate structure is an alternative FSD.

4.3.3 Stress, Displacement and Size Constraints

In this case, in addition to (4.18), we consider the displacement constraints

$$\sum_{i=1}^{n} \frac{N_i n_i^j l_i}{A_i E} \leqslant w_j^* \qquad j = 1, \ldots, p \tag{4.19}$$

as well as the size constraints

$$A_i \geqslant A_{i\min} . \tag{4.20}$$

The Lagrangian function can be written as

$$L(A_i, \lambda_j, \mu_i, \omega_i) = \sum_i k_i \, l_i \, A_i + \sum_j \lambda_j \left(\sum_i \frac{N_i n_i^j l_i}{A_i E} - w_j^* \right) +$$
$$+ \sum_i \mu_i \left(\frac{N_i}{A_i} - R_{ui} \right) + \sum_i \omega_i \left(A_{i\min} - A_i \right) . \tag{4.21}$$

When applied to (4.21), the optimality criteria give a system of equations which can be solved by iteration. Such iterative algorithms are proposed by several authors, e.g. Lipp (1976), Khan *et al.* (1979), Khot *et al.*(1979).

For a single displacement constraint combined with stress constraints Funaro (1974) has proposed a method which also allows for a constant limit stress value for compressed members. It is more advisable to determine first the FSD with suboptimized members and then, if the displacement constraint is not fulfilled, to apply the scaling operation of

$$A_i' = \frac{w_{\max}}{w^*} A_i . \tag{4.22}$$

The final A_i values are then obtained by using (4.17).

For problems with few variables ($n < 10$) the backtrack method (see Section 2.7) is also suitable.

4.4 Numerical Example

Consider the planar truss with parallel chords shown in Fig. 4.4 under the action of a uniformly distributed factored load $p = 25$ kN/m (the dead weight of approx. 1 kN/m being included). The load intensity for deflection calculations is $p' = 20$ kN/m. The truss is constructed from bars of square hollow cross-section made of steel 37. Limiting tensile stresses are as follows: for chord members $R_u = 190$ MPa; for other truss members (with due regard to the welded connections) $R_{u1} = 165$ MPa.

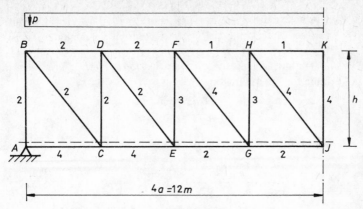

Fig. 4.4 – Numerical example of the optimum design of a planar truss with parallel chords (see Table 4.2)

The members are divided into 4 groups according to the numbering shown in Fig. 4.4. The profiles, suboptimized with regard to buckling for different bar-lengths and forces, can be obtained from Appendix A7. The suboptimization was carried out according to Section 3.2, by means of the backtrack method.

In the first part of the calculation, only *stress constraints* are considered. Thus FSD solutions are obtained for structures of different heights. Results, for active members only, are given in Table 4.2. It can be seen that the height $h = 1.25a$ gives the minimum volume.

With $E = 210$ GPa, the max. deflection is $w_{max} = 3.77$ cm. Let the allowable value be $w^* = 2.4$ cm. Considering this displacement constraint the optimum is

Table 4.2
Height Optimization under Stress Constraints of the Truss Shown in Fig. 4.4

h	Bar symbol	Group number (i)	Force N_i [kN]	A_i [cm^2]	Profile dimensions $b \times t$ [cm]	Half volume $V/2$ [cm^3]
a	KH	1	−600.0	36.0	18 × 0.5	
	JG	2	562.5	30.0	15 × 0.5	132 173
	EF	3	−112.5	12.0	10 × 0.3	
	EC	4	262.5	14.4	12 × 0.3	
$1.25a$	KH	1	−480.0	30.0	15 × 0.5	
	JG	2	450.0	24.0	15 × 0.4	118 226
	EF	3	−112.5	14.4	12 × 0.3	
	EC	4	210.0	11.2	14 × 0.2	
$1.50a$	KH	1	−400.0	28.0	14 × 0.5	
	AB	2	−262.5	24.0	15 × 0.4	125 515
	EF	3	−112.5	15.6	13 × 0.3	
	EC	4	175.0	9.6	12 × 0.2	

searched by two methods as follows: (1) hand calculations using the optimality criteria; (2) computations using the Fortran program for the backtrack method.

(1) Results obtained using the *optimality-criteria method* are given in Table 4.3. The formula for the scaling operation is

$$A_i' = \frac{3.77}{2.4} A_i = 1.5708 A_i .$$

The iteration is performed by using formula (4.17):

$$A_k^{(\nu+1)} = \frac{\sqrt{N_k n_k}}{E(w^* - w_{pass})} \sum_{m_p}' l_i \sqrt{N_i n_i} ; \quad w_{pass} = \sum_{m_c} \frac{N_i n_i l_i}{A_i^{(\nu)} E} .$$

Bars KH, JG, CE and FE are acitve, the others are passive. The calculations yield

$$A_k = \frac{2 \times 18\,646}{2.1 \times 10^4 (2.4 - 1.4697)} \sqrt{N_k n_{k_i}} = 1.9088 \sqrt{N_k n_k} .$$

This formula indicates that, for bar FE, $A_k = 12.8$ cm², while from the stress constraint we have obtained $A_k = 14.4$ cm². Thus, the bar becomes passive. A further iteration results in

$$A_k^{(2)} = 1.8720 \sqrt{N_k n_k} .$$

Table 4.3

Optimization with a Single Deflection Constraint of the Truss Shown in Fig. 4.4
(with Height $h = 1.25a$)

Bar sym-bol	Group num-ber	A_i [cm²]	$N_i =$ $N_i'/1.25$ [kN]	n_i	l_i [cm]	A_i' [cm²]	$A_k^{(1)}$ [cm²]	$A_k^{(2)}$ [cm²]	Profile dimensions $b \times t$ [cm]	A_k [cm²]
KH	1	30	−384	÷1.6	300	47.1	47.32	46.41	17 × 0.7	47.6
FH	1	30	−360	−1.2	300	47.1	47.32			
DF	2	24	−288	−0.8	300	47.7	39.66			
BD	2	24	−168	−0.4	300	37.7	39.66			
JG	2	24	360	1.2	300	37.7	39.66	39.59	20 × 0.5	40.0
EG	2	24	288	0.8	300	37.7	39.66			
CE	4	11.2	168	0.4	300	17.6	15.65	15.62	13 × 0.3	15.6
A.	2	24	−210	−0.5	375	37.7	39.66			
DC	2	24	−150	−0.5	375	37.7	39.66			
FE	3	14.4	− 90	−0.5	375	22.6	14.40	14.40	12 × 0.3	14.4
HG	3	14.4	− 30	−0.5	375	22.6	14.40			
BC	2	24	269	0.64	480.2	37.7	39.66			
DE	2	24	192	0.64	480.2	37.7	39.66			
FG	4	11.2	115	0.64	480.2	17.6	15.65			
HJ	4	11.2	38.4	0.64	480.2	17.6	15.65			
AC	4	11.2	0	0	300	17.6	15.65			
JK	4	11.2	0	0	375/2	17.6	15.65			

It can be seen that the change in the constant is small, so that further iteration is not required. With these final cross-sectional areas we get

$$w_{max} = 2.37 \text{ cm} < w^*$$

and $V/2 = 177\,043 \text{ cm}^3$.

(2) *Optimum design by the backtrack method*
The objective function is

$$V/2 = 600A_1 + 2910.4A_2 + 750A_3 + 1747.9A_4 \quad [\text{cm}^3].$$

Stress constraints: $A_1 \geqslant 30.0 \text{ cm}^2$;
$A_2 \geqslant 24.0 \text{ cm}^2$;
$A_3 \geqslant 14.4 \text{ cm}^2$;
$A_4 \geqslant 11.2 \text{ cm}^2$.

Deflection constraint:

$$\frac{313\,920}{A_1} + \frac{497\,245}{A_2} + \frac{22\,500}{A_3} + \frac{67\,326}{A_4} \leqslant \frac{2.4}{2}\, 21\,000 = 25\,200\, [\text{kN, cm}].$$

Series of discrete values for the variables A_i ($i = 1, \ldots, 4$) $[\text{cm}^2]$: $A_{i\min} = 15.0$; $A_{i\max} = 55.0$; $\Delta A_i = 2.5$.
Results: $A_1 = 55$; $A_2 = 35$; $A_3 = A_4 = 17.5 \text{ cm}^2$. Using suboptimized profiles we obtain $A_1 = 55.2$; $A_2 = 36.0$; $A_3 = A_4 = 17.6 \text{ cm}^2$. With these values we get $w_{max} = 2.34 \text{ cm}$ and $V/2 = 181\,857 \text{ cm}^3$.

Chapter 5

Statically Determinate Beams Subjected to Bending and Shear

In this chapter we consider the optimum design of homogeneous and hybrid I- and box girders, beams of circular and oval hollow cross-section, and sandwich beams. Since the effect of shear can usually be neglected, we first study beams subject predominantly to bending (Section 5.2); then, in Sections 5.3 and 5.4, the effect of shear is also considered. By neglecting shear and with the use of some approximations the minimum cross-sectional area design may be treated analytically. From the resulting simple formulae, comparative calculations can then be performed which should prove useful for designers. For more complicated problems, numerical minimization methods are used. Some problems of structural analysis, such as the ultimate shear strength of welded plate girders, and the statical and dynamical response of sandwich beams with flexurally stiff outer layers, are summarized in the Appendices.

5.1 Literature Survey for the Optimum Design of Welded Homogeneous I-Beams

Holt and Heithecker (1969) and Shive (1972) have studied the minimum cross-sectional area design of welded doubly symmetric I-beams on the basis of the Specification of the Amer. Inst. Steel Constr. The allowable compressive and shear stresses are given as functions of the beam sizes. These functions differ, depending on whether the elastic or plastic ranges are being considered. Therefore, the objective function of three variables should be minimized for more cases considering two stress constraints. The method of Lagrange multipliers is used resulting in an algebraic equation of higher order which can be solved numerically.

An optimized series of welded I-beams with web height/thickness ratios of 100, 150, 200 and 250 was computed by Leśniak (1970). The minimum cross-sectional area design was performed with the Monte-Carlo-method by imposing a stress constraint.

Annamalai (1970) also used the Specifications of AISC; however, he considered a cost function which expressed material and welding costs for an I-beam

with transverse stiffeners. The effect of a reduction in flange thickness using a
welded splice was also studied. The minimum cost design of simply supported,
arbitrarily loaded plate girders was performed by backtrack programming.

Bo *et al.* (1974) have studied the minimum cost design of welded doubly
symmetric I-beams with transverse and longitudinal stiffeners, considering the
costs of material, welding and painting. Simply supported, uniformly loaded
beams made of various steels were treated. A concentrated load at midspan was
also considered. The cost function of three variables was minimized by the
method of Lagrange multipliers considering a stress constraint. Checks on deflection and lateral buckling were performed subsequently.

The minimum cost design of welded I-beams with transverse stiffeners and a
welded splice at the flanges was solved using sequential linear programming by
Wills (1973). The design rules of BS 153 were applied, but the constraint on lateral buckling was not considered.

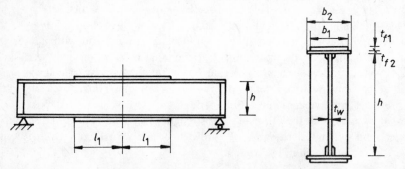

Fig. 5.1 – Welded plate girder optimized by Maeda and Konishi (1975)

The minimum cost design of the welded beam shown in Fig. 5.1 was treated
by Maeda and Konishi (1974–75). The objective function of eight variables, expressing the material and welding costs, was minimized by the sequential linear
programming method considering stress, deflection and size constraints. In addition to the 7 variables shown in Fig. 5.1 the type of steel was also treated as a
variable by taking into account steels with tensile strengths of 410, 500 and 580
MPa. The costs were expressed as functions of material, plate thickness and
width. Here, the Specifications for Steel Railway Bridge Design (Japan Society
of Civil Engineers) were used.

In another study Konishi and Maeda (1976) considered fabrication costs including costs of machining, shop welding, shop assembly and shop painting. I-
girders of constant web height and variable number of sections with different
flange dimensions have been optimized taking into account two different fabrication cost coefficients. The numerical values of optimum dimensions show that
the fabrication costs affect the number of different sections considerably. In the
case of high fabrication costs the total cost will be higher if the number of different sections is increased.

The same tendency is shown by Vachajitpan and Rockey (1977) in the case of an I-girder with transverse stiffeners. The optimum number of stiffeners decreases if the welding cost coefficient increases.

5.2 Optimum Design Neglecting Shear

5.2.1 Characteristics of Optimum Homogeneous I- and Box-Sections

In this section the minimum cross-sectional area design is treated considering the stress, deflection and local buckling constraints, but neglecting the dead weight. The limitation of web thickness is investigated in Section 5.2.4, and the cost of painting is taken into account in Section 5.2.5.

Welded doubly symmetric I- and box-sections, shown in Fig. 5.2, may usually be treated by means of the same formulae, by replacing h, t_w, b and t_f by h_1, t_{w1}, b_1 and t_{f1}, respectively. Special formulae for a box-section will be given only when they differ from those valid for an I-section.

(1) *The objective function* is

$$A = ht_w + 2bt_f \ . \tag{5.1}$$

(2) *The stress constraint* can be written as

$$M/W_x \leqslant R_u \tag{5.2}$$

where M is the factored maximum bending moment, W_x is the section modulus, and R_u is the limiting stress. Equation (5.2) may be expressed in the form

$$W_x \geqslant W_0 = M/R_u \ , \tag{5.3}$$

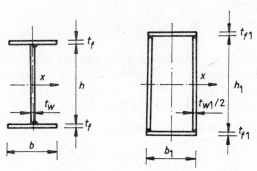

Fig. 5.2 – Dimensions of I- and box cross-sections

W_0 being the required section modulus. The approximate formula for the moment of inertia may be taken as

$$I_x = t_w h^3 / 12 + bt_f h^2 / 2 \tag{5.4}$$

and, with a further approximation, the section modulus can be expressed as

$$W_x = 2I_x/h = t_w h^2/6 + b t_f h. \tag{5.5}$$

(3) *The deflection constraint,* neglecting the effect of shear deformations, is given by

$$w_{max} = C_w/I_x \leqslant c^* L , \tag{5.6}$$

or, alternatively, in the form

$$I_x \geqslant I_0 = \frac{C_w}{c^* L}. \tag{5.7}$$

C_w is a constant depending on load, span length L and modulus of elasticity E. For instance, for a simply supported, uniformly loaded beam with constant cross-section its value is

$$C_w = \frac{5pL^4}{384E} \tag{5.8}$$

where p is the load intensity, c^* is the allowable deflection ratio, e. g. for simply supported beams in buildings $c^* = 1/100 - 1/300$ (see Section 1.2.1).

(4) By using the limit values of plate slenderness, *local buckling constraints* may be defined separately for web and flange. The problem of plate buckling is treated briefly in Appendix A3. Considering formulae (A3.11)–(A3.14) the local buckling constraint for the web plate of an I-beam may be given in the form

$$t_w/h \geqslant \beta , \tag{5.9a}$$

while, for a box beam,

$$t_{w1}/(2h_1) \geqslant \beta . \tag{5.9b}$$

For steels 37 with $R_{u\,37} = 200$ MPa β can be taken as $1/145$. For other stress levels this constant is given by

$$1/\beta = 145 \sqrt{R_{u\,37}/\sigma_{max}} , \tag{5.10}$$

and, for other types of steel with a limiting stress R_u,

$$\frac{1}{\beta} = 145 \sqrt{\frac{R_{u\,37}}{\sigma_{max}}} \sqrt{\frac{R_{u\,37}}{R_u}}. \tag{5.11}$$

Finally, for Al-alloys, the ratio E_{al}/E_s also affects the constant:

$$\frac{1}{\beta} = 145 \sqrt{\frac{R_{u\,37}}{\sigma_{max}}} \sqrt{\frac{R_{u\,37}}{R_{u.\,al}}} \sqrt{\frac{E_{al}}{E_s}}. \tag{5.12}$$

(5) The *local buckling constraint for the compressed flange* of an I-beam is given by

$$t_f/b \geqslant \delta \, , \qquad\qquad (5.13\text{a})$$

and for a box beam by

$$t_{f1}/b_1 \geqslant \delta_1 . \qquad\qquad (5.13\text{b})$$

In elastic design — see Table A3.1 (Appendix A3) — we have

$$\delta \cong \delta_1 = 1/30 \, . \qquad\qquad (5.14)$$

This value should be corrected if a different stress level and/or another type of steel are applicable, as was the case for the value of β. It is worth noting that other δ values can also be used because, as it will be seen in connection with formula (5.27), the optimum dimensions of web (h, t_w) and the optimum value for the cross-sectional area of the flanges $(A_f = bt_f)$ do not depend on δ.

(6) In the *optimum design procedure* the objective function of four variables (5.1) is to be minimized, considering constraints (5.3), (5.7), (5.9) and (5.13). It can be seen from (5.1) that the constraints due to local buckling (5.9) and (5.13) should be active, i. e. they should be treated as equalities; thus the number of variables is reduced to two.

In the case of an active deflection constraint the stress level is $\sigma_{max} < R_u$ and hence the β and δ values may be corrected. If this correction is neglected the problem can easily be solved graphoanalytically.

With the introduction of new variables

$$x_1 = h^2 \qquad\qquad (5.15)$$

$$x_2 = b^2 \qquad\qquad (5.16)$$

the objective function (5.1) may be written in the form

$$A = \beta h^2 + 2\delta b^2 = \beta x_1 + 2\delta x_2 . \qquad\qquad (5.17)$$

The new form of the stress constraint (5.3) is

$$W_x = \beta h^3/6 + \delta b^2 h = \beta x_1^{3/2}/6 + \delta x_1^{1/2}x_2 \geqslant W_0 \, . \qquad (5.18)$$

The equation of the limit curve for this stress constraint is thus

$$x_2^{(\sigma)} = \frac{W_0}{\delta x_1^{1/2}} - \frac{\beta}{6\delta}x_1 \, . \qquad\qquad (5.19)$$

The deflection constraint (5.7) can be expressed as

$$I_x = \frac{\beta h^4}{12} + \frac{\delta b^2 h^2}{2} = \frac{\beta x_1^2}{12} + \frac{\delta}{2}x_1 x_2 \geqslant I_0 \qquad\qquad (5.20)$$

so that the equation of the limit curve is

$$x_2^{(w)} = \frac{2I_0}{\delta x_1} - \frac{\beta}{6\delta}x_1. \qquad (5.21)$$

The abscissa x_{10} of the point of intersection of the two limit curves is then

$$x_{10} = 4I_0^2/W_0^2 . \qquad (5.22)$$

In the graphoanalytical method we need the inclination of the tangents for the limit curves at the point x_{10} (Fig. 5.3):

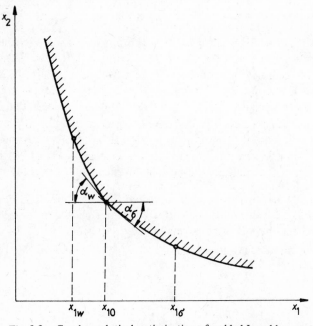

Fig. 5.3 – Graphoanalytical optimization of welded I- and box-sections

$$|\tan \alpha_w| = \left| \frac{dx_2^{(w)}}{dx_1} \right|_{x_1 = x_{10}} = \frac{W_0^4}{8\delta I_0^3} + \frac{\beta}{6\delta} .$$

Similarly, from (5.18) one obtains

$$|\tan \alpha_\sigma| = \frac{W_0^4}{16\delta I_0^3} + \frac{\beta}{6\delta} .$$

The inclination of the objective function (5.17) is

$$|\tan \alpha_A| = \beta/(2\delta) .$$

The stress constraint is active if $|\tan\alpha_A| \leqslant |\tan\alpha_\sigma|$, i. e. when

$$16\beta I_0^3 \leqslant 3W_0^4 \ . \tag{5.23}$$

On the other hand, the deflection constraint is active if $|\tan\alpha_A| \geqslant |\tan\alpha_w|$, i. e.

$$8\beta I_0^3 \geqslant 3W_0^4 \ . \tag{5.24}$$

Finally, in the case

$$8\beta I_0^3 \leqslant 3W_0^4 \leqslant 16\beta I_0^3 \tag{5.25}$$

the optimum is given by the point of intersection.

If the stress constraint is active, the optimum x_1-value may be obtained from

$$|\tan\alpha_A| = \left| \frac{dx_2^{(\sigma)}}{dx_1} \right|$$

$$x_1^{1/2}{}_{\mathrm{opt}} = h_\sigma = \sqrt[3]{\frac{3W_0}{2\beta}} \ . \tag{5.26}$$

If the deflection constraint is active, the approximate solution (for $\sqrt{\sigma_{\max}/R_u} = 1$) can be calculated from

$$|\tan\alpha_A| = \left| \frac{dx_2^{(w)}}{dx_1} \right|$$

$$x_1^{1/2}{}_{\mathrm{opt}} = h_w = \sqrt[4]{\frac{6I_0}{\beta}} \ . \tag{5.27}$$

Summarizing the three cases, we have:

(a) If

$$I_0^3 \leqslant \frac{3}{16\beta}\,W_0^4$$

the stress constraint is active; h_σ and b_σ can then be calculated through (5.26) and (5.19) respectively;

(b) If

$$I_0^3 \leqslant \frac{3}{8\beta}\,W_0^4 \leqslant 2I_0^3$$

the stress and deflection constraints are active; $h_{\mathrm{opt}} = 2I_0/W_0$, while b_{opt} may be obtained from (5.19) or (5.21);

(c) If

$$I_0^3 \geqslant \frac{3}{8\beta}\,W_0^4$$

Table 5.1

Characteristics of the Optimum Cross-Section

optimized with stress constraint		optimized with deflection constraint	
I-section	Box-section	I-section	Box-section
$h_\sigma = \sqrt[3]{1.5W_0/\beta}$	$h_{1\sigma} = \sqrt[3]{0.75W_0/\beta}$	$h_w = \sqrt[4]{6I_0/\beta}$	$h_{1w} = \sqrt[4]{3I_0/\beta}$
$t_{w\sigma} = \beta h_\sigma$	$t_{w1\sigma} = 2\beta h_{1\sigma}$	$t_{w.w} = \beta h_w$	$t_{w1.w} = 2\beta h_{1w}$
$A_\sigma = 2\beta h_\sigma^2 = \sqrt[3]{18\beta W_0^2}$	$A_{1\sigma} = 4\beta h_{1\sigma}^2 = \sqrt[3]{36\beta W_0^2}$	$A_w = 4\beta h_w^2/3 = \sqrt{32\beta I_0/3}$	$A_{1w} = 8\beta h_{1w}^2/3 = \sqrt{64\beta I_0/3}$
$t_{f\sigma} = h_\sigma\sqrt{\beta\delta/2}$	$t_{f1\sigma} = h_{1\sigma}\sqrt{\beta\delta}$	$t_{f.w} = h_w\sqrt{\beta\delta/6}$	$t_{f1.w} = h_{1w}\sqrt{\beta\delta/3}$
$b_\sigma = t_{f\sigma}/\delta$	$b_{1\sigma} = t_{f1\sigma}/\delta$	$b_w = t_{f.w}/\delta$	$b_{1w} = t_{f1}/\delta$
$I_{x\sigma} = \beta h_\sigma^4/3$	$I_{x1\sigma} = 2\beta h_{1\sigma}^4/3$	$I_{xw} = \beta h_w^4/6$	$I_{x1w} = \beta h_{1w}^4/3$
$W_{x\sigma} = 2\beta h_\sigma^3/3$	$W_{x1\sigma} = 4\beta h_{1\sigma}^3/3$	$W_{xw} = \beta h_w^3/3$	$W_{x1w} = 2\beta h_{1w}^3/3$

the deflection constraint is active; the approximate h_w and b_w can be calculated from (5.27) and (5.21) respectively.

Characteristics for the optimum cross-sections are given in Table 5.1.

If the deflection constraint is active, a more exact solution may be obtained by considering the corrected local buckling constraints as follows:

$$t_w = \beta h \sqrt{\sigma/R_u} \tag{5.28a}$$

$$t_f = \delta b \sqrt{\sigma/R_u} \,. \tag{5.28b}$$

Through the use of these formulae the section modulus can be written as

$$W_x = \sqrt{\frac{\sigma}{R_u}} \left(\frac{\beta h^3}{6} + \delta b^2 h \right) ,$$

while the max. stress is given by

$$\sigma = \frac{M}{W_x} = \sqrt{\frac{R_u}{\sigma}} \frac{M}{\beta h^3/6 + \delta b^2 h} \,.$$

Introducing the notation $C = \beta/6$ in the last equation we get

$$\sigma^{3/2} = \frac{\sqrt{R_u}\, M}{C h^3 + \delta b^2 h} \,.$$

Moreover, with the notation $x_3 = b/h$, the insertion of σ into (5.28) gives

$$t_w = 6C \sqrt[3]{\frac{W_0}{C + \delta x_3^2}} \tag{5.29a}$$

and

$$t_f = \delta x_3 \sqrt[3]{\frac{W_0}{C + \delta x_3^2}} \,. \tag{5.29b}$$

Then, on the basis of (5.1) and (5.29) one obtains

$$A = 2h(3C + \delta x_3^2) \sqrt[3]{\frac{W_0}{C + \delta x_3^2}} \,. \tag{5.30}$$

By using (5.4) and (5.29) we get

$$I_x = \frac{h^3}{2} W_0^{1/3} (C + \delta x_3^2)^{2/3} = I_0 \,,$$

from which

$$h = 2^{1/3} I_0^{1/3} W_0^{-1/9} (C + \delta x_3^2)^{-2/9} .$$ (5.31)

Inserting now (5.31) into (5.30) we obtain

$$A = 2^{4/3} I_0^{1/3} W_0^{2/9} (3C + \delta x_3^2)(C + \delta x_3^2)^{-5/9} .$$ (5.32)

The condition $dA/dx_3 = 0$ then gives

$$x_{3 \, \text{opt}} = \sqrt{\frac{3C}{2\delta}} = \sqrt{\frac{\beta}{4\delta}} .$$ (5.33)

By using (5.33) and (5.31) one obtains the more exact value

$$h_{we} = \sqrt[9]{\frac{32 I_0^3}{25 W_0 C^2}} = 1.53050 \sqrt[9]{\frac{I_0^3}{\beta^2 W_0}} .$$ (5.34)

From (5.33) we also get

$$b_{we} = h_{we} \sqrt{\frac{\beta}{4\delta}} .$$ (5.35)

Finally, on the basis of (5.32) and (5.33) the minimum area is

$$A_{\min} = 3.07371 \sqrt[9]{\beta^4 W_0^2 I_0^3} .$$ (5.36)

Note that in the case of a box-section, β in formulae should be replaced by 2β.

Numerical example. Data: steel 37, $R_u = 200$ MPa, $M = 2500$ kNm. Therefore, $W_0 = M/R_u = 12\,500$ cm^3; $\beta = 1/145$. Two cases for the deflection constraint should be considered: (a) $I_0 = 5 \times 10^5$ cm^4; (b) $I_0 = 12 \times 10^5$ cm^4.

Case (a): since $I_0 = 5 \times 10^5$ cm$^4 < \sqrt[3]{3 W_0^4/(16\beta)} = 8.723 \times 10^5$ cm^4, the stress constraint is active, so that $h_\sigma = 139.6$ cm from (5.26), and, by using Table 5.1, $A_{\min} = 268.7$ cm^2. For purposes of comparison, it is interesting to check the case when both stress and deflection constraints are active. Then $h = = 2I_0/W_0 = 80.0$ cm and

$$A = \frac{W_0^2}{I_0} + \frac{8\beta I_0^2}{3 W_0^2} = 341.9 \text{ cm}^2 .$$

This solution is not optimal and the difference is 27%.

Case (b): since

$$I_0 = 12 \times 10^5 \text{ cm}^4 > \sqrt[3]{3 W_0^4/(8\beta)} = 10.990 \times 10^5 \text{ cm}^4 ,$$

the deflection constraint is active, and the approximate solution, according to (5.27), is $h_w = 179.8$ cm, while, from Table 5.1, $A_{\min w} = 297.1$ cm^2. With the more exact expression (5.34) we obtain $h_{we} = 172.3$ cm and, from (5.36), $A_{\min we} = 291.0$ cm^2. The difference between the two values is thus only 2%.

5.2.2 Some Comparative Studies

By means of the simple formulae obtained in the previous section it is possible to compare the strength characteristics of various types of sections (Farkas 1969a). These comparisons are very useful for designers. In the accompanying calculations only the formulae of optimum sections obtained for the case of active stress constraint will be used.

5.2.2.1 Cross-Sectional Areas of I- and Box-Sections

From Table 5.1 we have

$$\frac{A_1 - A}{A} = \sqrt[3]{2} - 1 = 0.26 \,,$$

so that the area of a box-section is 26% larger than that of an I-section. Hence the latter is more economical for pure bending and should be used provided a larger torsional stiffness is not required and that lateral buckling is not a problem.

5.2.2.2 I-Beams Made of Steels 37 or 52

According to Table 5.1

$$A = \sqrt[3]{18\beta(M/R_u)^2} \,,$$

and from (A3.11) (Appendix A3)

$$\beta = \sqrt{\frac{12(1 - \nu^2)R_u}{k\pi^2 E}} \; .$$

Therefore, one obtains

$$A \sim R_u^{-1/2} \; .$$

In the case of static loads $R_{u\,52} = 1.4R_{u\,37}$ and

$$1 - \frac{A_{52}}{A_{37}} = 1 - \sqrt{\frac{R_{u\,37}}{R_{u\,52}}} = 0.15 \; .$$

Thus, the mass of an I-beam made of steel 52 is 15% less than that corresponding to steel 37. (It is worth noting at this stage that *hybrid* beams are more economical (see Section 5.2.8)). However, it should be pointed out that the *deflection* of a beam made of steel 52 is larger. Considering the formula

$$I_x = \frac{\beta h_\sigma^4}{3} = \frac{\beta}{3} \left(\frac{3M}{2\beta R_u} \right)^{4/3}$$

it can be seen that $I_x \sim R_u^{-3/2}$. Thus,

$$\frac{w_{52}}{w_{37}} - 1 = \frac{I_{x\,37}}{I_{x\,52}} - 1 = \left(\frac{R_{u\,52}}{R_{u\,37}} \right)^{3/2} - 1 = 0.66 \,,$$

i. e. the deflection is 66% larger.

5.2.2.3 I-Beams Made of Steel 37 or Al-Alloy

Formulae used in Section 5.2.2.2 show that $A \sim R_u^{-1/2}$ and $A \sim E^{-1/6}$ so that the ratio of masses is

$$1 - \frac{G_{al}}{G_s} = 1 - \frac{\rho_{al}}{\rho_s} \sqrt{\frac{R_{us}}{R_{ual}}} \sqrt[6]{\frac{E_s}{E_{al}}} \,.$$

With values of $R_{ual} = 150$ MPa, $R_{us} = 200$ MPa, $\rho_s = 7850$, $\rho_{al} = 2750$ kg/m^3, and $E_{al}/E_s = 1/3$ we get

$$1 - G_{al}/G_s = 0.51 \,,$$

i. e. the mass of an I-beam made of Al-alloy is 51% less than that made of steel 37. The *deflections* are, however, larger. Considering that

$$\beta \sim E^{-1/2} R_u^{1/2} \qquad \text{and} \qquad I_x \sim E^{1/6} R_u^{-3/2}$$

one obtains

$$\frac{w_{al}}{w_s} = \frac{E_s I_s}{E_{al} I_{al}} = \left(\frac{E_s}{E_{al}} \right)^{7/6} \left(\frac{R_{ual}}{R_{us}} \right)^{3/2} = 2.34 \,,$$

i. e. the deflections of an I-beam made of Al-alloy are 134% larger.

5.2.2.4 Torsional Stiffnesses of I- and Box-Sections

The torsional inertia of an I-section may be calculated as

$$I_t = \frac{\alpha_0}{3} (h t_w^3 + 2 b t_f^3) = \frac{\alpha_0}{3} h^4 \beta^2 \left(\beta + \frac{\delta}{2} \right).$$

With the values $\alpha_0 = 1.5$, $\beta = 1/140$, $\delta = 1/30$, we obtain $I_t = 6.07 \times 10^{-7} h^4$.

The torsional inertia of a box-section, on the other hand, is given by

$$I_{t1} = \frac{4b_1^2 h_1^2}{2b_1/t_{f1} + 4h_1/t_{w1}} = \frac{2\beta h_1^4}{1 + \delta/\beta} = 2.52 \times 10^{-3} h_1^4 \ .$$

Since $h_1 = 2^{-1/3}h$, we get $I_{t1}/I_t = 1647$, i. e. the torsional stiffness of a box-section is much larger than that of an I-section.

5.2.2.5 Normal Stresses in I- and Box-Beams Due to Warping Torsion

For a simply supported I-beam (Fig. 5.4) the maximum warping torsional moment is given by

$$B_{\omega max} = Fe \, \frac{\tanh \alpha l}{\alpha} ; \qquad \alpha^2 = \frac{GI_t}{EI_\omega}$$

where the warping section constant is

$$I_\omega = \frac{h^2 b^3 t_f}{24} = 1.59 \times 10^{-5} h^6.$$

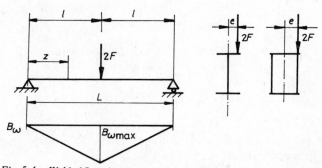

Fig. 5.4 – Welded I- or box-beam subjected to bending and torsional actions

With $G/E = 0.8/2.1$ and $I_t = 6.07 \times 10^{-7} h^4$ we get $\alpha^2 = 1.45 \times 10^{-2} h^{-2}$ and $\alpha l = 0.120 \, l/h$. If one further assumes that $\alpha l < 0.5$ (i. e. $h > 0.24 l$) the following approximations can be used: $\tanh \alpha l \approx \alpha l$ and $B_{\omega max} \cong Fel$.

The maximum normal warping stress is given by

$$\sigma_{\omega max} = \frac{B_{\omega max}}{I_\omega} \, \omega_{max} = \frac{6Fel}{hb^2 t_f}$$

whilst the maximum bending stress is

$$\sigma_{b max} = \frac{Fl}{W_x} = \frac{3Fl}{2\beta h^3} \ .$$

The ratio of these two stresses is then

$$\frac{\sigma_{\omega\max}}{\sigma_{b\max}} = \frac{8e}{b}$$

which shows that, if $e > b/80$, $\sigma_{\omega\max}/\sigma_{b\max} > 0.1$, i. e. that *the I-beam is very sensitive to warping torsion.*

In the case of a box-beam (see, e. g., (Panc 1959)) the following expressions apply:

$$\sigma_{\bar{\omega}\max} = \frac{B_{\bar{\omega}\max}}{I_{\bar{\omega}}} \, \bar{\omega}_{\max}$$

$$B_{\bar{\omega}\max} = \mu Fe \, \frac{\tanh\alpha_1 l}{\alpha_1} \; ; \quad \alpha_1^2 = \frac{\mu G I_{t1}}{EI_{\bar{\omega}}} \; ; \quad \mu = 1 - \frac{I_{t1}}{I_p}$$

$$I_p = \oint_A r_C \, dA = h_1 t_{w1}(b_1/2)^2 + 2b_1 t_{f1}(h_1/2)^2 = 4.337 \times 10^{-3} h_1^4$$

$$I_{t1} = 2.521 \times 10^{-3} h_1^4 \; ; \qquad \mu = 0.41868 \, .$$

For the corner point of a box-section one obtains

$$\bar{\omega}_{\max} = \frac{h_1 b_1}{4} - \frac{2h_1 b_1}{2b_1/t_{f1} + 4h_1/t_{w1}}\left(\frac{h_1}{t_{w1}}\right) = -7.488 \times 10^{-2} h_1^2$$

$$I_{\bar{\omega}} = \bar{\omega}_{\max}^2 \left(\frac{2}{3}b_1 t_{f1} + \frac{1}{3}h_1 t_{w1}\right) = 5.34 \times 10^{-5} h_1^6 \; .$$

Thus, $\alpha_1 = 2.744/h_1$. If $\alpha_1 l > 1.5$, i. e. if $h_1 < 1.83l$, the following approximations may be used: $\tanh\alpha_1 l \approx 1$, and

$$|B_{\bar{\omega}\max}| = \mu Fe/\alpha_1 = 0.15258 Feh_1 \, .$$

The maximum warping and bending stresses are

$$\sigma_{\bar{\omega}\max} = 213.95 \, Fe/h_1^3 \; ; \qquad \sigma_{b\max} = 105 \, Fl/h_1^3$$

and thus their ratio is $\sigma_{\bar{\omega}\max}/\sigma_{b\max} = 2.04 \, e/l$. This ratio is larger than 10% if $e > 0.05l$ and hence it can be concluded that *box-beams are insensitive* to warping torsion.

5.2.2.6 Lateral Buckling of I- and Box-Beams

Consider the I- and box-beams shown in Fig. 5.4. In the case of the I-beam, the elastic critical lateral buckling moment, according to Vol'mir (1967), is

$$M_{cr} = \frac{4.23}{L} \sqrt{1 + \frac{\pi^2}{\alpha^2 L^2}} \sqrt{EI_y GI_t}$$

where

$$I_y = \frac{2b^3 t_f}{12} = \frac{\delta}{6} h^4 \left(\frac{b}{h}\right)^4 = \frac{\beta^2 h^4}{24\delta} = 0.6377 \times 10^{-4} h^4 \ .$$

If $\alpha l < 0.5$, i. e. if $\alpha L < 1$, we can use the following approximation:

$$\sqrt{1 + \pi^2/(\alpha^2 L^2)} \approx \pi/(\alpha L)$$

so that

$$M_{cr} = \frac{4.23\pi^2}{L^2} EI_y \sqrt{\frac{I_\omega}{I_y}} \ .$$

Moreover, as shown in Section 5.2.2.5, $I_\omega = 1.59 \times 10^{-5} \ h^6$, and hence

$$M_{cr} = 2.63 \times 10^{-3} Eh^5/L^2 \ .$$

We use the condition

$$M_{cr}/W_x \leqslant R_{ub}$$

where $W_x = 2\beta h^3/3$ and $R_{ub} \leqslant R_u$ is the limiting stress for overall buckling. With values of $R_{ub} = R_u = 190$ MPa and $E = 210$ GPa we obtain the following condition for lateral buckling

$$L \leqslant h \sqrt{\frac{3 \times 2.63 \times 10^{-3} E}{2\beta R_u}} \cong 25h$$

from which it is clear that *the I-beam is sensitive to lateral buckling.*

In the case of a box-beam the expression for the elastic critical lateral buckling moment is

$$M_{cr1} = \frac{4.23}{L} \sqrt{1 + \frac{\pi^2}{\alpha_1^2 L^2}} \sqrt{EI_{y1} GI_{t1}} \ .$$

On the basis of our discussion in Section 5.2.2.5, $\sqrt{1 + \pi^2/(\alpha_1^2 L^2)} \approx 1$. Furthermore,

$$I_{y1} = \frac{b_1^3 t_{f1}}{6} + \frac{h_1 t_{w1} b_1^2}{4} = \frac{2\beta^2 h_1^4}{3\delta} = 1.02 \times 10^{-3} h_1^4$$

and, with $I_{t1} = 2.521 \times 10^{-3} h_1^4$ and $G/E = 0.8/2.1$, we obtain

$$M_{cr1} = \frac{4.23 E I_{y1}}{L} \sqrt{\frac{G I_{t1}}{E I_{y1}}} = \frac{4.1874 \times 10^{-3} E}{L} h_1^4 .$$

With $W_x = 4\beta h_1^3/3$ the condition $M_{cr1}/W_x \leqslant R_u$ gives

$$L \leqslant \frac{3 \times 4.1874 \times 10^{-3} E}{4\beta R_u} h_1 \cong 486 h_1 ,$$

and hence we can conclude that *lateral instability of box-beams* in practice *will not*, in general, *occur*.

5.2.3 Limitations Imposed by the Web Thickness

From the point of view of production technology a size constraint should be specified for the web thickness, i. e.

$$t_w \geqslant t_0 \tag{5.37a}$$

or, in the case of a box-section

$$t_{w1}/2 \geqslant t_{01} . \tag{5.37b}$$

If the stress constraint (5.3) is active and, if, from (5.26), we obtain a thickness

$$t_w = \beta h_\sigma < t_0 ,$$

then the calculation should be repeated on the assumption that, instead of (5.9), the size constraint (5.37) is active. Inserting the expression for bt_f from (5.1) into (5.5) one obtains

$$W_x = \frac{Ah}{2} - \frac{t_0 h^2}{3} = W_0 ,$$

from which we get

$$A = \frac{2 W_0}{h} + \frac{2 t_0 h}{3} . \tag{5.38}$$

Finally, the condition $dA/dh = 0$ yields

$$h_{opt} = \sqrt{3 W_0/t_0} . \tag{5.39}$$

5.2.4 The Effect of Dead Weight

For a beam of constant cross-section the maximum bending moment can be considered as consisting of two parts as follows:

$$M = M_1 + mA \ .$$

M_1 is the maximum bending moment due to live load, while mA represents the effect of dead weight. If the stress constraint (5.3) is active, we can insert the expression for bt_f from (5.1) into (5.5), so as to obtain

$$W_x = \frac{Ah}{2} - \frac{t_w h^2}{3} = W_0 = \frac{M}{R_u} = W_1 + a_0 A \tag{5.40}$$

where $W_1 = M_1/R_u$ and $a_0 = m/R_u$. If the web buckling constraint (5.9a) is active, expression (5.40) yields

$$A = \frac{2W_1}{h - 2a_0} + \frac{2\beta h^3}{3(h - 2a_0)} \tag{5.41}$$

in the case of an I-section. The condition $dA/dh = 0$ gives

$$h^2 (h - 3a_0) = \frac{3W_1}{2\beta} \ . \tag{5.42}$$

This equation can be solved by the trial and error method, a suitable starting value for h being that obtained from (5.26). Note that in the case of box-sections β should be replaced by 2β in the above calculations.

5.2.5 Effect of Painting Costs

In the objective function, the cost of surface preparation and painting could be considered in addition to the cost of materials. For an I-beam we obtain

$$K = K_m + K_p = k_m \rho A + k_p (2h + 4b) \ . \tag{5.43}$$

If the stress constraint (5.3) is active, Eq. (5.41) takes the form

$$A = \frac{2W_0}{h} + \frac{2\beta h^2}{3} \tag{5.44}$$

when the dead weight is neglected ($a_0 = 0$). Furthermore, with the approximation $b = 0.3h$, (5.43) may be written as

$$\frac{K}{k_m \rho} = \frac{2W_0}{h} + \frac{2\beta h^2}{3} + \frac{3.2 k_p h}{k_m \rho} \ . \tag{5.45}$$

The problem is illustrated by means of the data used for the numerical example of Section 5.2.1. Thus, $W_0 = 1.25 \times 10^7$ mm^3 and $\beta = 1/145$; we also adopt $k_m = 0.3$ \$/kg, $\rho = 7850$ kg/m^3, $k_p = 4.4$ \$/m^2 (see, e. g., (Bo *et al.* 1974) or (Wills 1973)). Expression (5.45) then reduces to

$$\frac{K}{k_m\rho} = \frac{2.5 \times 10^7}{h} + \frac{h^2}{2.175 \times 10^5} + 6.0h \quad [\text{mm}^2]. \tag{5.46}$$

Condition $dK/dh = 0$ yields $h_{opt} = 1209$ mm, as against $h_{opt} = 1396$ mm for the case when painting costs are not taken into account.

By means of this example it can be shown that *the sensitivity* of the cost function (5.46) is less than that of the simpler objective function (5.44): if, instead of h_{opt}, we were to adopt $1.2h_{opt}$, A and K would experience increases of 3.5% and 2.8% respectively.

5.2.6 Beams of Circular Hollow Section

Spiral welded pipes can be used as girders for various purposes, e. g. galleries of belt conveyors, self-supporting pipelines, etc. (see (Bahke 1970), (Golubenko *et al.* 1975)). The optimum design of circular hollow sections was first treated by Shanley (1960). Subsequently, the design for minimum cross-sectional area of aluminium tubes subjected to combined bending and torsion was solved by Felton and Dobbs (1967).

With the notation $\delta_c = D/t$, the cross-sectional area of a thin-walled tube (Fig. 5.5) can be expressed as

$$A = D\pi t = D^2\pi/\delta_c . \tag{5.47}$$

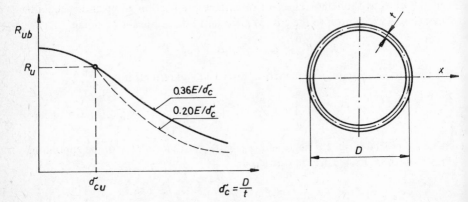

Fig. 5.5 – Approximate local buckling diagram for a circular tube subjected to bending

The moment of inertia is

$$I_x = \frac{D^3 \pi t}{8} = \frac{D^4 \pi}{8\delta_c} \tag{5.48}$$

and the section modulus is given by

$$W_x = \frac{D^2 \pi t}{4} = \frac{D^3 \pi}{4\delta_c} . \tag{5.49}$$

The stress constraint is expressed as

$$\sigma_{max} = \frac{M}{W_x} \leqslant R_{ub} \tag{5.50}$$

where M is the maximum bending moment and R_{ub} is the limiting stress governed by local buckling of the tube. According to Vol'mir (1967), the local buckling strength for bending conditions can be calculated with the formula applicable to axially compressed cylindrical shells. In the elastic zone this gives

$$R_{ub} = 0.36E/\delta_c ,$$

while, allowing for the effect of initial imperfections and residual welding stresses, the following formula may be proposed:

$$R_{ub} = kE/\delta_c , \quad k = 0.20. \tag{5.51}$$

In the plastic zone we take $R_{ub} = R_u$ (see Fig. 5.5), due to lack of experimental data. It should be noted that the critical buckling stress values proposed by the ECCS Recommendations (Vandepitte and Rathé 1980) for axially loaded circular cylindrical shells are much larger than those proposed here.

The critical stress of *ovalization* may be calculated on the basis of Reissner and Weinitschke (1963):

$$\sigma_{cr} = 0.46E/\delta_c .$$

It can be seen that Eq. (5.51) gives a safety factor of 2.3 against ovalization.

In the optimum design procedure the cross-sectional area should be minimized considering the stress and deflection constraints. This two-dimensional problem can be treated graphoanalytically in the coordinate system $D^2 - \delta_c$. On the basis of (5.49), (5.50) and (5.51) *the stress constraint* can be expressed as

$$\delta_c \leqslant \sqrt{\frac{kE\pi D^3}{4M}} = \sqrt{\frac{kE\pi}{4M}} (D^2)^{3/4} \quad \text{for} \quad \delta_c \geqslant \delta_{cu} = \frac{kE}{R_u}. \tag{5.52}$$

Similarly, through the use of (5.7) and (5.48), *the deflection constraint* may be written in the form

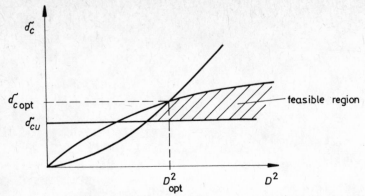

Fig. 5.6 – Graphoanalytical optimization of a beam of circular hollow section

$$\delta_c \leqslant \frac{\pi}{8I_0}(D^2)^2; \qquad I_0 = \frac{C_w}{c^*L} . \tag{5.53}$$

Figure 5.6 shows the feasible region. Since the contours of the objective function are straight lines passing through the origin of coordinates, the optimum is given by the intersection point of that limit curves of the two constraints, provided that $\delta_{copt} \geqslant \delta_{cu}$. On the basis of (5.52) and (5.53) we get

$$D_{opt} = \left(\frac{16I_0^2 kE}{\pi M}\right)^{1/5} ; \qquad \delta_{copt} = \frac{\pi}{8I_0}\left(\frac{16I_0^2 kE}{\pi M}\right)^{4/5} . \tag{5.54}$$

If $\delta_{c\,opt} < \delta_{cu}$ (Fig. 5.7),

$$\delta'_{copt} = \delta_{cu} , \tag{5.55}$$

and from the active stress constraint one obtains

$$D'_{opt} = \sqrt[3]{\frac{4M\delta_{cu}^2}{kE\pi}} = \sqrt[3]{\frac{4MkE}{\pi R_u^2}} . \tag{5.56}$$

Then the minimum cross-sectional area is given by

$$A_0 = \sqrt[3]{\frac{16M^2\pi}{kER_u}} . \tag{5.57}$$

It is worth comparing the circular shape with the box-section. By using the formula given in Table 5.1 we get

$$\frac{A_{1\sigma}}{A_0} = \sqrt[3]{\frac{9\beta kE}{4\pi R_u}} .$$

With the values of $\beta = 1/145, k = 0.2, E = 210$ GPa and $R_u = 200$ MPa one obtains $A_{1\sigma}/A_0 = 1.01$, thus one can conclude that the circular hollow section

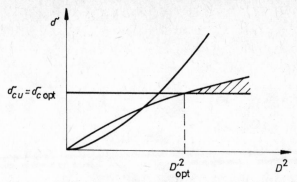

Fig. 5.7 — A further case of the optimization treated in Fig. 5.6

may be as economical as the box-section, provided that the required dimensions are available. Note that a box beam needs four longitudinal welds and its shape factor relating to wind pressure is greater than that of a circular tube; on the other hand, circular tubes need stiffening rings at the points of action of concentrated load, mainly above supports.

With the above values we obtain $\delta'_{\rho opt} = 210$; in the case of an existing thickness limitation, $t \geqslant t_0 = 4$ mm, we get $D_{opt} = 840$ mm.

Finally, we mention that the minimum weight design of stiffened cylindrical shells subject to bending is treated by Patel and Patel (1980), who use both the SUMT and Box-methods.

5.2.7 Beams of Oval Hollow Section

The thin-walled oval hollow section constructed from two vertical web plates and two semi-circular flanges (Fig. 5.8) is more economical than the circular one. It is used, for example, for crane girders (Peters 1973). Experimental results relat-

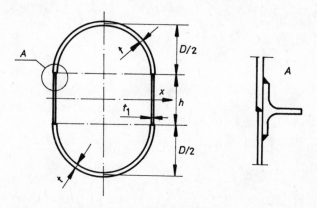

Fig. 5.8 — Thin-walled oval-like cross-section

ing to the local buckling behaviour of such beams are not to be found in the literature. However, if longitudinal stiffeners are used at the junctions of webs and flanges, one could assume the local buckling formula for webs to be applicable, while Eq. (5.51) might be used for the compressed semi-circular flange.

The cross-sectional area is given by

$$A = 2ht_1 + D\pi t \tag{5.58}$$

while the moment of inertia may be written as

$$I_x = \frac{h^3 t_1}{6} + \frac{D^3 \pi t}{8} + D^2 h t + \frac{D\pi h^2 t}{4} . \tag{5.59}$$

With the notation $\vartheta_c = D/h$, *the local web buckling* constraint can be expressed as

$$t_1 \geqslant \beta h \sqrt{\frac{\sigma_w}{R_u}} = \frac{\beta h}{\sqrt{1 + \vartheta_c}} \quad ; \tag{5.60}$$

here σ_w is the bending stress at the fibre $y = h/2$. If only the stress constraint were to be considered, *the local flange buckling constraint* could be written as

$$t \geqslant D/\delta_{cu} , \tag{5.61}$$

δ_{cu} being calculated from (5.52). With constraints (5.60) and (5.61) both active, the objective function (5.58) may be expressed as

$$A = 2\beta h^2 \left[(1 + \vartheta_c)^{-1/2} + \frac{\pi \vartheta_c^2}{2\beta \delta_{cu}} \right]. \tag{5.62}$$

By using (5.59) we obtain the following expression for *the stress constraint:*

$$W_x = \frac{2I_x}{h + D} = \frac{\beta h^3}{3} \left[(1 + \vartheta_c)^{-3/2} + \frac{3\vartheta_c^2 \pi}{4\beta \delta_{cu}(1 + \vartheta_c)} \left(2 + \frac{8}{\pi} \vartheta_c + \vartheta_c^2 \right) \right] \geqslant$$

$$\geqslant W_0 = \frac{M}{R_u} . \tag{5.63}$$

Solving for h in (5.63), and inserting the resulting expression into (5.62), we get

$$A = 2 \times 3^{2/3} \beta^{1/3} W_0^{2/3} f(\vartheta_c) , \tag{5.64}$$

where

$$f(\vartheta_c) = \frac{(1 + \vartheta_c)^{-1/2} + \pi \vartheta_c^2/(2\beta \delta_{cu})}{\left[(1 + \vartheta_c)^{-3/2} + \dfrac{3\vartheta_c^2 \pi}{4\beta \delta_{cu}(1 + \vartheta_c)} \left(2 + \dfrac{8}{\pi} \vartheta_c + \vartheta_c^2 \right) \right]^{2/3}} .$$

The minimization of $f(\vartheta_c)$ can be performed by a one-dimensional numerical search procedure. In this way, for $\beta = 1/140$ and $\delta_{cu} = 210$, one obtains $\vartheta_{copt} = 1.7$, $f_{min} = 0.57912$, $A_{min} = 0.464\ W_0^{2/3}$. The economy of oval sections can be demonstrated by comparing this expression with the A_0 obtained for the circular hollow section (5.57):

$$A_0 = \sqrt[3]{\frac{16\pi R_u}{kE}}\ W_0^{2/3} = 0.621\ W_0^{2/3}\ .$$

This economy may be achieved only if the local buckling constraints are active. In the case of the thickness limitation $t_{1\,min} = 3$ mm, we get

$$h = \frac{t_{1\,min}\sqrt{1 + \vartheta_{copt}}}{\beta} = 690\ \text{mm},$$

so that $D = \vartheta_{copt}\,h \vcentcolon 170$ mm, $t = D/\delta_{cu} = 5.6$ mm, $W_x = 1.23 \times 10^7$ mm^3. A detailed numerical example is given in (Farkas 1972).

5.2.8 Hybrid I-Beams

An economical way of using high strength steels is the fabrication of I-beams with flanges having a higher strength than the web (on the use of high strength steels see Section 1.1). Surveys on this topic were given by Kato and Okumura (1976), Kunihiro *et al.* (1976), Maeda *et al.* (1976). The static behaviour was treated by Maeda and Kawai (1972), while fatigue problems were investigated by Toprac and Natarajan (1971), and Yamasaki *et al.* (1976). Lateral buckling of hybrid I-beams was discussed by Nethercot (1976), design rules having been elaborated earlier by the ASCE (Design 1968). Several papers deal with the optimum design of hybrid girders, e. g. (Farkas 1969a), (Klinov 1973), (Schilling 1974), (Chong 1976), (Rondal and Maquoi 1977), (Sikalo 1977), (Kunitskiy 1978), (Farkas 1980c, Farkas and Szabó 1980).

5.2.8.1 Characteristics of the Cross-Section Optimized on the Basis of a Stress Constraint

The stress distribution of a hybrid I-beam in the limit state is shown in Fig. 5.9. Thus

$$N = (R_{uf} - R_{yw})h't_w/4 = R_{uf}(1 - \xi_h^2)ht_w/4 \tag{5.65}$$

because, with the notation $\xi_h = R_{yw}/R_{uf}$,

$$\frac{h'}{h} = \frac{R_{uf} - R_{yw}}{R_{uf}} = 1 - \xi_h\ . \tag{5.66}$$

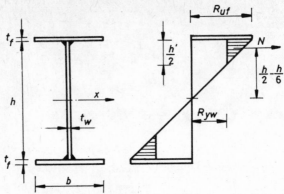

Fig. 5.9 – Limit state stress distribution in a hybrid I-beam subjected to bending

The limiting bending moment is approximately

$$M_u = R_{uf}\left(\frac{h^2 t_w}{6} + b t_f h\right) - 2N\left(\frac{h}{2} - \frac{h'}{6}\right).$$ (5.67)

On the basis of (5.65) and (5.67), and neglecting the dead weight, *the stress constraint* takes the form

$$\frac{M_u}{R_{uf}} = b t_f h + \frac{h^2 t_w}{12}(3\xi_h - \xi_h^3) \geqslant W_0.$$ (5.68)

The cross-sectional area is

$$A = h t_w + 2 b t_f$$ (5.69)

so that the area of the flanges is

$$2 b t_f = A - h t_w.$$ (5.70)

Finally, inserting (5.70) into (5.68), the stress constraint may be expressed as

$$A \geqslant \frac{2W_0}{h} + \frac{h t_w}{6}(6 - 3\xi_h + \xi_h^3).$$ (5.71)

The local buckling constraint of an unstiffened web is (see Section 5.2.1)

$$t_w/h \geqslant \beta;$$ (5.72)

for 37-steels β can be taken as $1/145$.

Treating both constraints (5.71) and (5.72) as active, we get

$$A = \frac{2W_0}{h} + \frac{\beta h^2}{6}(6 - 3\xi_h + \xi_h^3).$$ (5.73)

The material cost per unit length of a beam is

$$K_m = k_w \rho h t_w + 2 k_f \rho b t_f \tag{5.74}$$

where k_w and k_f are the material cost factors for the web and the flanges, respectively.

Introducing the notation $\eta_h = k_w/k_f$, and using (5.70) and (5.73), we obtain

$$\frac{K_m}{k_f \rho} = \frac{2 W_0}{h} + \frac{\beta h^2}{6} (6\eta_h - 3\xi_h + \xi_h^3). \tag{5.75}$$

The condition $dK_m/dh = 0$ gives

$$h_{\text{opt}} = \sqrt[3]{\frac{6 W_0}{\beta(6\eta_h - 3\xi_h + \xi_h^3)}} \tag{5.76}$$

which, for a homogeneous beam ($\eta_h = \xi_h = 1$), reduces to (5.26).

On the basis of (5.75) and (5.76), *the minimum cost* can be expressed as

$$\frac{K_{m\,\text{min}}}{k_f \rho} = \sqrt[3]{4.5 \, W_0^2 \beta(6\eta_h - 3\xi_h + \xi_h^3)}. \tag{5.77}$$

By using (5.73) and (5.76) one obtains the following cross-sectional area for the beam corresponding to minimum cost:

$$A_{K\,\text{min}} = \frac{\beta h^2}{2} (4\eta_h + 2 - 3\xi_h + \xi_h^3). \tag{5.78}$$

From (5.70) and (5.78) the area of the flanges is then simply

$$2 b t_f = A - \beta h^2 = \frac{\beta h^2}{2} (4\eta_h - 3\xi_h + \xi_h^3). \tag{5.79}$$

The local buckling constraint of the compressed flange (see Section 5.2.1) is

$$\frac{t_f}{b} \geqslant \delta = \frac{1}{30} \sqrt{\frac{R_{uf}}{R_{u\,37}}} \tag{5.80}$$

and can be regarded as active. Thus, from (5.79) we obtain

$$b = \frac{\beta h}{2} (4\eta_h - 3\xi_h + \xi_h^3)$$

and

$$t_f = \delta b = \frac{h}{2} \sqrt{\beta \delta (4\eta_h - 3\xi_h + \xi_h^3)}. \tag{5.81}$$

Finally, the approximate formula for the moment of inertia corresponding to the minimum cost solution can be written as

$$I_x = \frac{h^3 t_w}{12} + \frac{bt_f h^2}{2} = \frac{\beta h^4}{24} \left(12\eta_h + 2 - 9\xi_h + 3\xi_h^3\right) =$$

$$= (12\eta_h + 2 - 9\xi_h + 3\xi_h^3) \sqrt[3]{\frac{3}{32\beta} \left(\frac{M}{R_{uf}}\right)^4 \frac{1}{(6\eta_h - 3\xi_h + 3\xi_h^3)^4}} . \quad (5.82)$$

5.2.8.2 Characteristics of the Cross-Section Optimized on the Basis of a Deflection Constraint

If the deformations of the beam are limited to the elastic range, the deflection constraint is given by

$$I_x = \frac{h^3 t_w}{12} + \frac{bt_f h^2}{2} \geqslant I_0 . \quad (5.83)$$

By using (5.70), Eq. (5.83) can be rewritten in the form

$$A \geqslant \frac{4I_0}{h^2} + \frac{2ht_w}{3} . \quad (5.84)$$

If the deflection constraint is active, the actual maximum stress in the web (σ) is less than R_{uw} so that the exact web buckling constraint would be

$$t_w/h \geqslant \beta\sqrt{\sigma/R_{uw}} .$$

Instead of this, however, we use (5.72). According to (5.74) and (5.84), the material cost can be expressed as

$$\frac{K_m'}{k_f \rho} = \frac{4I_0}{h^2} + \frac{\beta h^2}{3} (3\eta_h - 1) . \quad (5.85)$$

The condition $dK_m'/dh = 0$ then gives

$$h_{opt}' = \sqrt[4]{\frac{12I_0}{\beta(3\eta_h - 1)}} . \quad (5.86)$$

From (5.84) and (5.86) the cross-sectional area is

$$A_{K\,min}' = \frac{\beta h^2}{3} (3\eta_h + 1) = (3\eta_h + 1) \sqrt{\frac{4I_0\beta}{3(3\eta_h - 1)}} \quad (5.87)$$

The area of the flanges is now

$$2bt_f = \frac{\beta h^2}{3} (3\eta_h - 2) \quad (5.88)$$

and the minimum cost may be expressed as

$$\frac{K'_{m\,min}}{k_f \rho} = \sqrt{\frac{16}{3}(3\eta_h - 1)\beta I_0} \; . \tag{5.89}$$

5.2.8.3 Some Comparative Calculations

The economy of hybrid I-beams can be demonstrated by means of the approximate formulae derived in the previous section. In our calculations we use the following two hybrid beams (stresses in MPa):

(a) 52/37-beam: $R_{uw} = 200$; $R_{yw} = 240$; $R_{uf} = 280$; $\xi_h = 0.86$; $\eta_h = 0.901$;

(b) 70/37-beam: $R_{uw} = 210$; $R_{yw} = 230$; $R_{uf} = 530$; $\xi_h = 0.434$; $\eta_h = 0.617$.

(1) *Comparison of weights* of hybrid and homogeneous beams optimized on the basis of the stress constraint. By using (5.78) and (5.76) we get

$$\frac{A_{hyb}}{A_{hom}} = \frac{4\eta_h + 2 - 3\xi_h + \xi_h^3}{\sqrt[3]{4(6\eta_h - 3\xi_h + \xi_h^3)^2}} \left(\frac{R_{uw}}{R_{uf}}\right)^{2/3} . \tag{5.90}$$

For 52/37 and 70/37 beams, (5.90) gives the values 0.805 and 0.602 respectively; thus, *hybrid beams are 20–40% lighter* than their homogeneous counterparts.

(2) *Comparison of material costs.* From (5.77) one obtains

$$\frac{K_{m\,hyb}}{K_{m\,hom}} = \frac{1}{\eta_h}\left(\frac{R_{uw}}{R_{uf}}\right)^{2/3} \sqrt[3]{\frac{6\eta_h - 3\xi_h + \xi_h^3}{4}} . \tag{5.91}$$

For 52/37 and 70/37 beams, (5.91) gives the values 0.845 and 0.746 respectively; thus, *hybrid I-beams result in 15–25% savings in material costs.*

(3) *Comparison of deflections.* Assuming that $\beta_{hyb} = \beta_{hom}$, and according to (5.82), we obtain

$$\frac{w_{hyb}}{w_{hom}} = \frac{I_{x\,hom}}{I_{x\,hyb}} = \frac{\sqrt[3]{2}\,(6\eta_h - 3\xi_h + \xi_h^3)^{4/3}}{12\eta_h + 2 - 9\xi_h + 3\xi_h^3}\left(\frac{R_{uf}}{R_{uw}}\right)^{4/3} . \tag{5.92}$$

For 52/37 and 70/37 beams, (5.92) gives the values 1.48 and 2.53 respectively; thus, *the deflections of hybrid beams are larger* than those of homogeneous girders. Therefore it is advisable to examine whether the cross-section optimized on the basis of a stress constraint does in fact satisfy the deflection constraint, i. e. whether $I_x \geqslant I_0$. I_x can be calculated from (5.82).

(4) *Comparison of material costs* for hybrid and homogeneous beams optimized on the basis of the deflection constraint. From (5.89) we obtain

$$\frac{K'_{m\,hyb}}{K_{m\,hom}} \cong \frac{1}{\eta_h}\sqrt{\frac{3\eta_h - 1}{2}} . \tag{5.93}$$

For 52/37 and 70/37 beams, (5.93) gives the values 1.024 and 1.057 respectively; thus, *if the deflection constraint is active, hybrid beams are uneconomical.*

5.2.8.4 Numerical Examples Solved by Backtrack Programming

If a more intricate objective function is defined and the dead weight is also taken into account, it is impossible to derive simple formulae for the optimum sections. Thus, the results obtained by means of the computer method of backtrack for three numerical examples will now be given (Farkas and Szabó 1980). The geometry of the simply supported, uniformly loaded, welded I-beam used in our investigation is shown in Fig. 5.10. The list of discrete values used in the calculations appears in Table 5.2. The lowering of flange thickness with a welded splice defined by the distance L_1 is also taken into consideration.

Table 5.2
Discrete Values Used in Calculations
(dimensions in cm)

Dimension	min	max	Δ	Number of values
h	80	120	5	9
b	10	42	2	17
t_w	0.4	2.0	0.1	17
t_f	0.8	2.4	0.2	9
L_1	$0.1L$	$0.3L$	$0.025L$	9

The distance of the vertical stiffeners is $N_s = L/a - 1$, where $a = 1.4h$. N_s is an integer (rounded off), and denotes the number of intermediate stiffeners. The thicknesses of the intermediate and end stiffeners are $0.8t_w$ and t_w, respectively. These stiffeners are attached to the web and the flanges by means of double fillet welds with dimensions $0.4\,t_w$ and $0.5\,t_w$, respectively. The material of the stiffeners is the same as that of the web. In the case of a symmetrical welded splice on the flanges $n_{s1} = L_1/a$ is an integer which has been rounded off.

The objective function expresses the costs of material, welding and painting:

$$K = K_m + K_W + K_p \tag{5.94}$$

where

$$K_m = k_w \rho h t_w L + 4 k_f \rho b_1 t_{f1} L_1 + 2 k_f \rho b t_f (L - 2L_1) +$$
$$+ 2 k_w \rho h t_w b_1 + 1.6 k_w N_{s1} \rho h t_w b_1 + 0.8 k_w \rho (N_s - N_{s1}) h t_w b$$
$$K_W = 2 k_W t_w L + 4 k_W t_w (h + b_1) + 3.2 k_W N_{s1} t_w (h + b_1) +$$
$$+ 1.6 k_W (N_s - 2N_{s1}) t_w (h + b) + 4 k_{W1} b_1 t_{f1}$$

$$K_p = k_p(2h + 4b)(L - 2L_1) + 2k_p(2h + 4b_1)L_1 + 2(N_s - 2N_{s1})k_p hb +$$
$$+ 4(1 + N_{s1})k_p hb_1 \ .$$

Note that indices w and W denote web and welding, respectively. The specific costs are as follows: *welding:* $k_W = 97.14\ \$/m^2$ (fillet welds), $k_{W1} = 62.86\ \$/m^2$ (butt welds in splices); *painting:* $k_p = 1.543\ \$/m^2$. It should be noted that, according to the European Recommendations (1978), the welding costs may be

Table 5.3
Limiting and Yield Stress and Material Cost Parameter for Steels Used in Calculations
(stresses in MPa)

Material	R_u	R_y	k_w, k_f [$\$/kg$]
37	200	240	0.391
52	280	400	0.434
85	550	750	0.634

considerably higher, because one man hour (the whole shop labour cost) may be equivalent to 30–50 kg steel. Steels 37, 52 and 85 are considered with the relevant data given in Table 5.3. The density is $\rho = 7850\ kg/m^3$. For beams without splices $L_1 = N_{s1} = k_{W1} = 0$ and $b_1 = b$ in (5.94).

The constraint due to maximum bending stress as well as those for local buckling of the web and the compressed flange are considered. Checks for the shear are also performed. However, the constraints of lateral buckling and deflection are not taken into account, nor is the design of vertical stiffeners allowed for.

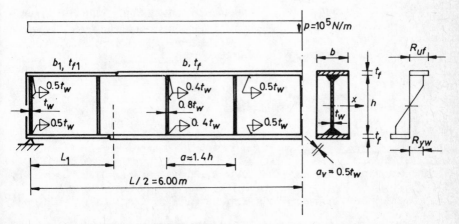

Fig. 5.10 – Simply supported welded I-beam with one welded splice on the flanges

Table 5.4
Data for Beams Without Splices Optimized on the Basis of Total Cost
(dimensions in cm)

Materials	h	t_w	b	t_f	Total cost [1000 $]	Savings [%]
37 homogeneous	120	0.9	30	2.0	1.017	0
52/37 hybrid	120	0.9	16	2.4	0.846	17
85/37 hybrid	120	0.9	12	1.4	0.714	30

The calculations are first carried out for beams without splices and the results are summarized in Table 5.4. It can be seen that hybrid beams result in 17–30% savings in the total cost. The homogeneous beams made of steel 52 or 85 are less economical than the hybrid ones. For example, the homogeneous beam made of steel 52 gives only 10% savings in the overall cost.

Table 5.5
Data for Beams Without Splices Optimized on the Basis of Material Cost
(dimensions in cm)

Materials	h	t_w	b	t_f	Total cost [1000 $]	Total cost from Table 5.4	Savings [%]
37 homogeneous	120	0.9	42	1.4	1.063	1.017	4.3
52/37 hybrid	120	0.9	24	1.6	0.880	0.846	3.9

In Table 5.5 the data of beams without splices optimized on the basis of material cost only are summarized. It may be seen that the consideration of a more realistic cost function can result in savings of about 4%.

The optimization of beams with splices is carried out in the following manner. With the data for the optimum beam without splices as a starting point, we take

Table 5.6
Data for Beams with a Flange Splice Optimized on the Basis of Total Cost
(dimensions in cm)

Materials	h	t_w	b	t_f	b_1	t_{f_2}	L_1	Total cost [1000 $]	Total cost from Table 5.4	Savings [%]
37 homogeneous	120	0.9	30	2.0	12	1.8	180	0.903	1.017	11.2
52/37 hybrid	120	0.9	16	2.4	10	1.4	210	0.760	0.846	10.1
85/37 hybrid	120	0.9	12	1.4	10	0.8	240	0.669	0.714	6.4

the distance of the splice L_1 (see Table 5.2) and calculate the bending moment at this cross-section. Knowing h, t_w and M_{max} we search b_1 and t_{f1} by means of the backtrack method taking into account the total cost function. The calculation should be carried out for all L_1-values. Clearly, the optimum L_1-value will be that for which the total cost is a minimum.

Data for beams with a flange splice are summarized in Table 5.6. It may be seen that the application of splices can result in savings of about 6–11%.

5.3 Optimum Design Considering the Effect of Shear

The effect of shear is mainly significant for short girders and for sandwich beams. Shear stresses occur mainly in webs of welded I- or box beams and lower their capability to resist buckling. Optimum design can be performed in two ways: (1) allowing for the shear by formulae of bending-shear interaction; (2) prescribing limitations for which the shear can be neglected. Some results relating to the ultimate shear strength of plate girders are summarized in Appendix A3.3.

5.3.1 Optimum Design of Welded Homogeneous I- and Box-Beams Subjected to Bending and Shear

The objective function is given by (5.1). *The stress constraint* can be expressed as

$$\sqrt{\sigma_M^2 + 3\tau^2} \leqslant R_u \ . \tag{5.95}$$

The web buckling constraint for an I-beam is (see (A3.14))

$$\frac{h}{t_w} \leqslant \frac{1}{\beta} = 145 \sqrt[4]{\frac{1 + 3(\tau/\sigma_M)^2}{1 + 20(\tau/\sigma_M)^2}} \tag{5.96a}$$

while for box-beams (see Fig. 5.2) it is given by

$$\frac{2h_1}{t_{w1}} \leqslant \frac{1}{\beta} \ . \tag{5.96b}$$

In Eqs (5.96) $\sigma_M = M/W_x$; $\tau = Q/(ht_w)$;

$$W_x = \left[\frac{h^3 t_w}{12} + \frac{bt_f}{2}(h + t_f)^2 \right] \frac{1}{\frac{h}{2} + t_f} \ .$$

The buckling constraint for the compressed flange is given by (5.13).

Limitation of the web thickness:

$$t_w \geqslant t_0 \qquad \text{(I-beams)} \tag{5.97a}$$

$$t_{w1}/2 \geqslant t_0 \qquad \text{(box-beams)}. \tag{5.97b}$$

Technological constraint: in order to enable a suitable welding of the longitudinal junctions between the webs and flanges, it is necessary to formulate an appropriate criterion, as mentioned in Section 1.2.2:

$$t_f \geqslant \alpha_w t_w ; \qquad \alpha_w = 0.5 - 1 \qquad \text{(I-section)} \tag{5.98a}$$

$$t_{f1} \geqslant \alpha_w t_{w1}/2 \qquad \text{(box-section)} . \tag{5.98b}$$

The above optimization problem was treated in (Farkas 1978c), being solved for box-sections by the SUMT method.

A suboptimized series of box cross-sections subjected to bending and shear was computed by the backtrack method for the following data: $\alpha_w = 0.7$; $t_0 = 3$ mm; $R_u = 200$ MPa; $h_0 = 100$ cm (for some values of the parameters, $m = 100 \, M/(R_u h_0^3)$ and $q = Qh_0/(2\,M)$). The following series of discrete values for the variables were used (dimensions in cm):

	min	max	Δ
h_1	40	296	2
b_1	10	138	2
$t_{w1}/2$	0.3	3.5	0.1
t_{f1}	0.4	5.2	0.2

The results appear in Table 5.7.

5.3.2 $M - Q$ Interaction Diagrams for Transversely Stiffened Hybrid and Homogeneous I-Beams Optimized for Bending

$M - Q$ interaction diagrams can be determined for optimum I-sections by using the Rockey − Porter − Evans-method treated in Appendix A3.3, with resulting formulae (A3.17) − (A3.26).

(1) *Homogeneous I-girder* (Fig. 5.11) optimized on the basis of the stress constraint. The data are as follows (stresses in MPa): $h/t_w = 145$; steel 37; $R_y = 240$; $R_u = 200$; $\tau_y = 0.577 R_y = 138.5$; $\xi_h = \eta_h = 1$; distance between stiffeners $b = 1.5h$; $\delta = 1/30$.

In can be seen from the interaction diagram that, in a cross-section subjected to $M_{\max} = 0.7407 M_p$ ($\sigma_{\max} = R_u$), the effect of shear can be neglected if

$$\tau \leqslant \tau_{uB} = \frac{Q_B \tau_{yw}}{Q_{yw} \gamma_b} = 0.6440 \; \frac{138.5}{1.08} = 82.6 \text{ MPa} .$$

For cross-sections loaded predominantly in shear the following criterion can be prescribed:

$$\tau \leqslant \tau_{uS} = \frac{Q_C \tau_{yw}}{Q_{yw} \gamma_b} = 93.6 \text{ MPa} .$$

Table 5.7
Data of Optimized Box Sections Subjected to Bending and Shear
(dimensions in cm, areas in cm^2);
$m = 100 \; M/(R_u h_0^3); q = Q h_0/(2M); h_0 = 100 \text{ cm}; R_u = 200 \text{ MPa}$

m	q	h_1	$t_{w1}/2$	$A_f = b_1 t_{f1}$	A
	0.10	56	0.4	10.8	66.4
	0.20	52	0.4	12.8	67.2
0.10	0.30	48	0.4	16.0	70.4
	0.40	58	0.5	8.4	74.8
	0.50	54	0.5	10.8	75.6
	0.10	66	0.5	36.0	138.0
	0.20	74	0.6	28.0	144.8
0.30	0.30	88	0.7	16.0	155.2
	0.40	78	0.7	25.6	160.4
	0.50	70	0.7	35.2	168.4
	0.10	110	0.8	64.8	305.6
	0.20	110	0.9	66.0	330.0
1.00	0.30	114	1.0	62.4	352.8
	0.40	124	1.1	50.4	373.6
	0.50	132	1.2	39.6	396.0
	0.10	174	1.3	104.0	660.4
	0.20	176	1.5	100.8	729.6
3.00	0.30	182	1.6	100.8	784.0
	0.40	186	1.8	96.8	863.2
	0.50	212	2.0	42.0	932.0
	0.10	270	2.1	208.0	1550.0
10.0	0.20	294	2.5	145.6	1761.2
	0.30	294	2.8	172.8	1992.0

(2) *Hybrid I-girder* (Fig. 5.12) optimized on the basis of the stress constraint with the following data (stresses in MPa): flanges made of steel 52, web of 37, so that $R_{yw} = 240$ and $R_{uf} = 280, R_{yf} = 360; \xi_h = 0.86; \eta_h = 0.901; \delta = 1/25$.

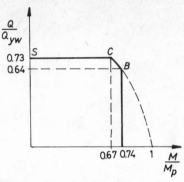

Fig. 5.11 – $M-Q$ interaction diagram for homogeneous I-beams optimized for bending;
steel 37; $h/t_w = 145$; $b/t_f = 30$; $a/h = 1.5$; $\tau_{uS} = 93.6$ MPa; $\tau_{uB} = 82.6$ MPa

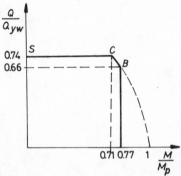

Fig. 5.12 – $M-Q$ interaction diagram for hybrid I-beams optimized for bending;
steels 52/37; $h/t_w = 145$; $b/t_f = 25$; $a/h = 1.5$; $\tau_{uS} = 94.9$ MPa; $\tau_{uB} = 84.7$ MPa

The interaction diagram shows that, in a cross-section acted upon by $M = 0.7715M_p$ ($\sigma_{max} = R_{uf}$), the effect of shear can be neglected if

$$\tau \leqslant \tau_{uB} = \frac{Q_B \tau_{yw}}{Q_{yw} \gamma_b} = 0.6607 \frac{138.5}{1.08} = 84.7 \text{ MPa} .$$

For cross-sections under mainly shear loading, the shear criterion is given by

$$\tau \leqslant \tau_{uS} = \frac{Q_C \tau_{yw}}{Q_{yw} \gamma_b} = 94.9 \text{ MPa} .$$

5.4 Sandwich Beams with Outer Layers of Box Cross-Section

The static and dynamic response of such beams is summarized in Appendix A4.
An optimum design problem for the symmetrical sandwich beam shown in Fig.
5.13 can be defined as follows (Farkas and Jármai 1982).

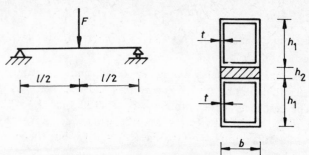

Fig. 5.13 – Symmetrical sandwich beam with flexurally stiff outer layers

The objective function may be expressed as the total material cost

$$K_m = 2A_1 lk_1 + bh_2 lk_2 + 2blk_3$$

where k_1, k_2 and k_3 are the material cost factors for the face profile of cross-sectional area A_1 (e. g. Al-alloy rectangular tube), damping layer (e. g. rubber or polyurethan foam) and adhesive, respectively.

The following constraints may be considered.

(1) *Normal stress constraint* in the extreme fibre of the outer layers:

$$\sigma_{1\,max} \leqslant R_{u1} .$$

$\sigma_{1\,max}$ can be calculated by formulae (A4.2) – (A4.8) given in the Appendix.

(2) *Constraint of the shear stress* in the core:

$$\tau_2 \leqslant \tau_{u2} .$$

The formula for τ_2 is given by (A4.7).

(3) *Deflection constraint:*

$$w_{max} \leqslant w^* .$$

w_{max} is to be calculated by (A4.1), w^* being the allowable deflection.

In the case of a forced vibration, the dynamic force is, according to (A4.16), $F_2 \cong F_1/\eta_d$, F_1 being the applied force and η_d the damping (loss) factor calculated by means of formulae (A4.12) – (A4.15).

(4) *Local buckling constraint* for the compressed flange:

$$t/b \geqslant \delta .$$

δ may be determined by (A3.10), given in Appendix A3.

(5) *Local buckling constraint* for webs subjected to combined bending and compression:

$$t/h_1 \geqslant \beta.$$

β can be calculated by means of (A3.10) – (A3.14), with the approximation $\sigma_N/\sigma_M \cong 1$.

Numerical example. Consider a simply supported sandwich beam of span $l =$ $= 300$ cm, constructed from two Al-alloy tubes and a rubber layer glued between them. The cost factors are as follows: $k_1 = 4400\ \$/m^3$; $k_2 = 1000\ \$/m^3$; $k_3 = 20\ \$/m^2$. $R_{u1} = 100$ MPa; $\tau_{u2} = 2.5$ MPa; $w^* = 2$ cm; $E_{a1} = 70$ GPa; for rubber, $G_s = 2.36$ MPa and $G_d = 7.0$ MPa, $\eta_{d2} = 0.18$. For the local buckling constraints we take

$$\frac{1}{\beta} = 145 \sqrt[4]{\frac{4}{174}} \sqrt{\frac{2}{3}} \cong 40$$

and

$$\frac{1}{\delta} = 30 \sqrt{\frac{200}{100}} \sqrt{\frac{1}{3}} \cong 25 .$$

Furthermore, $F_1 = 475$ N, and stresses and deflections should be calculated with the dynamic force $F_2 = F_1/\eta_d$. The list of discrete values for the variables is as follows: (dimensions in mm):

	min	max	Δ
h_1	40	200	10
h_2	5	50	10
b	40	200	10
t	2	10	1

The computations were performed by means of the backtrack method.

Results: optimum dimensions (in mm): $h_1 = 120$; $b = 50$; $t = 3$; $h_2 = 15$. Furthermore, $\eta_d = 0.0535$; $F_2 = 8878$ N, $\sigma_{1\,max} = 95$ MPa; $\tau_2 = 0.237$ MPa, $w_{max} = 1.72$ cm (the deflection constraint was not active); $K_{m\,min} = 37.36\ \$$.

In order to illustrate the economy of sandwich beams of this type, let us compare the above optimum beam with a homogeneous one assuming that the material damping factor of a homogeneous Al-alloy box girder is $\eta_{dh} = \eta_d/6$.

Considering Eq. (A4.16), the beam should be designed for a dynamic force $F = F_2\eta_d/\eta_{dh} = 53.268$ N, with both stress and local web buckling constraints. For web buckling we take

$$\frac{1}{\beta'} = 145 \sqrt{\frac{200}{R_u}} \sqrt{\frac{E_{al}}{E_s}} = 145 \sqrt{\frac{2}{3}} = 118 .$$

The max. bending moment is $M = 53.268 \times 3/4 = 39.95$ kNm, and thus the required section modulus is $W_0 = M/R_u = 399.5$ cm^3. According to Table 5.1

$$h_{opt} = \sqrt[3]{0.75\, W_0/\beta'} = 32.82 \text{ cm};$$

$t_w/2 = \beta' h_{opt} = 0.278$ cm, $t_f = h_{opt}\sqrt{\beta'\delta} = 0.6$ cm. With rounded values of $h = 350$, $t_w/2 = 3$, $t_f = 6$ and $b_f = 135$ mm one obtains $A = 3720$ mm^2. The material cost is $K_m = 49.10$ $ which is 31% higher than that of the sandwich beam.

The characteristics of the sandwich beam and the homogeneous box beam are summarized in Fig. 5.14 and Table 5.8.

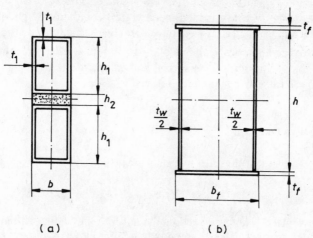

(a)   (b)

Fig. 5.14 – Optimum cross-sections (dimensions in mm):
(a) sandwich beam, $h_1 = 120$, $h_2 = 15$, $b = 50$, $t_1 = 3$;
(b) homogeneous box-beam, $h = 350$, $t_w/2 = 3$, $b = 135$, $t_f = 6$

Table 5.8
Comparison of the Characteristics of the two Optimized Beams

Characteristics	Sandwich beam Fig. 5.14 (a)	Homogeneous box-beam Fig. 5.14 (b)
Maximum dynamic stress $\sigma_{1\,max}$ [MPa]	95.0	99.4
Maximum dynamic deflection w_{max} [mm]	17.2	5.9
Loss factor η_d	0.0535	0.0089
Material costs [$]	37.36	49.10

5.5 Optimum Design of Crane Runway Girders

5.5.1 General Aspects of Design

In heavy-duty crane and crane runway girders with high loading cycles fatigue cracks have been observed in the vicinity of the top flange-web junction. The cracks occur in the web at the intermediate stiffeners and in the web-to-flange weld between stiffeners. The cracking is caused by high local compression and bending stresses introduced by eccentrically acting wheel loads.

To avoid these cracks, some improving methods may be used. It is proposed, for example, to apply a thicker flange, a thicker upper part of the web, a stronger rail, a resilient pad under the rail, a box-section instead of an open I-beam, etc. In the design of heavy loaded metallurgical crane runway girders the local compression and bending stresses due to wheel loads should be considered.

The following numerical example of a simply supported asymmetrical I-girder illustrates the application of the special constraints in the optimum design process relating to the above mentioned effects.

5.5.2 Objective Function

In the optimum design the cross-sectional area (Fig. 5.15)

$$A = b(t_{f1} + t_{f2}) + \frac{h}{3}(t_{w1} + 2t_{w2}) \tag{5.99}$$

is to be minimized. The cross-sectional area of a longitudinal stiffener and that of the rail is neglected. The six unknowns are as follows: b, h, t_{w1}, t_{w2}, t_{f1} and t_{f2}.

5.5.3 Stress Constraint for the Lower Flange at Midspan

This constraint is related to the bending stresses:

$$\frac{\psi_d M_{max}}{W_{x2}} \leqslant \frac{R_f}{\gamma_f} \quad \text{or} \quad \frac{M_0 + \psi_d M_{max}}{W_{x2}} \leqslant \frac{R_y}{\gamma} \tag{5.100}$$

where M_0 and M_{max} are the maximum bending moments due to dead and live load, respectively, ψ_d is the dynamic factor, R_f is the fatigue strength of the welded structural part, γ_f is the safety factor, W_{x2} is the section modulus given by the following formulae:

$$W_{x2} = \frac{I_x}{\frac{2h}{3} - y_S + \frac{t_{f2}}{2}} \tag{5.101}$$

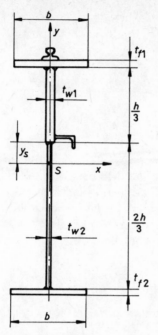

Fig. 5.15 – Cross-section of the crane runway girder

$$y_S = \frac{1}{A}\left[t_{w2}\frac{2h^2}{9} + t_{f2}b\left(\frac{t_{f2}}{2} + \frac{2h}{3} \right) - t_{w1}\frac{h^2}{18} - t_{f1}b\left(\frac{t_{f1}}{2} + \frac{h}{3} \right) \right] \quad (5.102)$$

$$I_x = \frac{t_{w1}h^3}{324} + t_{w1}\frac{h}{3}\left(\frac{h}{6} + y_S \right)^2 + \frac{2t_{w2}h^3}{81} + t_{w2}\frac{2h}{3}\left(\frac{h}{3} - y_S \right)^2 +$$

$$+ bt_{f1}\left(\frac{h}{3} + y_S + \frac{t_{f1}}{2} \right)^2 + bt_{f2}\left(\frac{2h}{3} - y_S + \frac{t_{f2}}{2} \right)^2 . \quad (5.103)$$

In the numerical example a runway girder for two casting cranes is considered with span length of $L = 24$ m. The distribution of the wheel load is shown in Fig. 5.16.

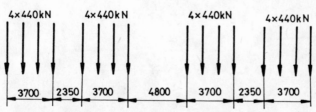

Fig. 5.16 – Wheel loads from two cranes

The maximum bending moments are as follows:

$$\psi_d M_{max} = 1.3 \times 1696.6 \times 10^7 = 2.2056 \times 10^{10} \text{ Nmm}$$

and

$$M_0 = \frac{1.1 p_a L^2}{8} = \frac{1.1 \times 7.85 \times 10^{-5} \times 24000^2}{8} (A + A_r) =$$

$$= 6.2172 \times 10^3 (A + 15044) \text{ [Nmm]}. \qquad (5.104)$$

$A_r = 15044$ mm^2 is the cross-sectional area of the rail KS 120. According to the Swiss standard SIA 161, the yield stress of the steel Fe 360 is $R_y = 235$ MPa, for thicknesses of $40 \leqslant t \leqslant 60$ mm $R_y = 215$ MPa, $R_y/\gamma = 134$ MPa.

For the fatigue class B (continuous longitudinal welds and machined transverse welds), for a number of load cycles $N = 500000$ the fatigue strength is $R_f = 183$ MPa and the allowable stress for plate thicknesses $40 \leqslant t \leqslant 60$ mm is

$$\frac{R_f}{\gamma_f} \times \frac{215}{235} = \frac{183 \times 215}{1.25 \times 235} = 134 \text{ MPa} .$$

Thus, in the numerical example, this stress constraint can be expressed as

$$\sigma_{z2} = \frac{1}{W_{x2}} [6.2172 \times 10^3 (A + 15044) + 2.2056 \times 10^{10}] \leqslant 134 \text{ MPa} .(5.105)$$

5.5.4 Shear Stress Constraints at the Support

According to the SIA 161, these constraints have the following form:

$$\gamma \tau_{max} = \frac{1}{A_w} \gamma \left(\frac{1.1 p_a L}{2} + \psi_d Q_{max} \right) \leqslant$$

$$\leqslant \begin{cases} \tau_y = \dfrac{R_y}{\sqrt{3}} = \dfrac{235}{\sqrt{3}} = 135 \text{ MPa} & (5.106a) \\\\ 0.9 \sqrt{\tau_{cr} \tau_y} & (5.106b) \end{cases}$$

where

$$A_w = \frac{h t_{w1}}{3} + \frac{2 h t_{w2}}{3} \qquad (5.107)$$

and, for

$$\frac{a}{h} \leqslant 1$$

$$\tau_{cr} = \frac{k_\tau \pi^2 E}{12(1-\nu^2)} \left(\frac{t_{w2}}{h}\right)^2 = \frac{1.89 \times 10^5}{(h/t_{w2})^2} \left[4 + \frac{5.34}{(a/h)^2}\right]. \quad (5.108)$$

The effect of the longitudinal stiffener is neglected. For a safety factor of $\gamma = 1.6$, and a distance of transverse stiffeners $a = 2000$ mm, the shear stress constraints are given by

$$\frac{1}{A_w}\left[1.0362(A + 15044) + 4.3706 \times 10^6\right] \leqslant$$

$$\leqslant \begin{cases} \dfrac{135}{1.6} = 84.4 \text{ MPa} & (5.109a) \\[4mm] \dfrac{2841.3}{h/t_{w2}}\sqrt{4 + \dfrac{5.34}{(a/h)^2}}. & (5.109b) \end{cases}$$

5.5.5 Stress Constraint for the Upper Flange at Midspan

According to the SIA 161, in this constraint the effects of the horizontal load are also taken into account:

$$\frac{1}{W_{x1}}\left(\frac{1.1 p_a L^2}{8} + 0.9\, \psi_d M_{\max}\right) + 0.7\frac{M_y}{W_y} + 0.7\frac{N_z}{A_f} \leqslant \frac{R_y}{\gamma} \quad (5.110)$$

where

$$W_{x1} = \frac{I_x}{\dfrac{h}{3} + y_s + \dfrac{t_{f1}}{2}}\, ; \qquad W_y = \frac{b^2 t_{f1}}{6}\, ; \qquad A_f = b t_{f1}\, .$$

N_z is the compression force in the upper flange due to bending of the horizontal bracing.

In the numerical example, this constraint takes the following form:

$$\frac{1}{W_{x1}}\left[6.2172 \times 10^3 (A + 15044) + 1.98502 \times 10^{10}\right] + \frac{40.824 \times 10^7}{b^2 t_{f1}} +$$

$$+ \frac{5.103 \times 10^5}{b t_{f1}} \leqslant 134 \text{ MPa}. \quad (5.111)$$

5.5.6 Fatigue Constraint for the Upper Web-to-Flange Weld

According to the Soviet research results incorporated in the Soviet building code SNiP II–23–81, the local effects due to the wheel load should be considered by the following constraint:

$$0.5 \sqrt{\sigma_{z1}^2 + 0.36\tau^2} + 0.4\sigma_{y1} + 0.5\sigma_{y2} \leqslant 75 \text{ MPa} \qquad (5.112)$$

where σ_{z1} and τ are the normal and shear stresses, respectively, in the upper fibre of the web from the main loads, σ_{y1} is the compression stress due to the wheel load:

$$\sigma_{y1} = \alpha_0 \frac{0.3\,\psi_d F}{t_{w1}} \sqrt[3]{\frac{t_{w1}}{1.15I_f + I_r}} \, . \qquad (5.113)$$

If a resilient pad is used under the rail, $\alpha_0 = 0.75$. Furthermore, $I_f = bt_{f1}^3/12$; I_r is the moment of inertia of the rail, for rails KS 120 $I_r = 3.817 \times 10^7$ mm^4.

The normal stress due to the local bending (or torsion), according to Beleňa and Neždanov (1979), may be calculated with the following formula

$$\sigma_{y2} = \frac{1.33 M_t t_{w1} a}{I_t h} \, . \qquad (5.114)$$

In the SNiP II–23–81 the following simplified formula is proposed:

$$\sigma_{y2} = \frac{2M_t t_{w1}}{I_t} \qquad (5.115)$$

where

$$M_t = eF + \frac{0.75 h_r F}{10} \qquad (5.116)$$

and

$$I_t = \frac{bt_{f1}^3}{3} + I_{tr} \, . $$

For a rail KS 120 $h_r = 170$ mm, $I_{tr} = 1.31 \times 10^7$ mm^4.

In the numerical example, the eccentricity of the vertical load $F = 4.4 \times 10^5$ N is $e = 15$ mm, and the following two constraints should be considered:
at midspan ($\tau \approx 0$):

$$0.5\sigma_{z1} + 0.4\sigma_{y1} + 0.5\sigma_{y2} \leqslant 75 \text{ MPa} \qquad (5.117)$$

at supports ($\sigma_{z1} \approx 0$):

$$0.3\tau_{max} + 0.4\sigma_{y1} + 0.5\sigma_{y2} \leqslant 75 \text{ MPa} \qquad (5.118)$$

where

$$\sigma_{z1} = \frac{1}{W_{x1}}[6.2172 \times 10^3 (A + 15044) + 2.2056 \times 10^{10}] \quad (5.119)$$

and

$$\sigma_{y1} = \frac{1.287 \times 10^5}{t_{w1}} \sqrt[3]{\frac{t_{w1}}{\dfrac{1.15\,bt_{f1}^3}{12} + 3.817 \times 10^7}} \qquad (5.120)$$

5.5.7 Other Constraints

The web buckling constraint may be calculated by SNiP II–23–81 taking into account the longitudinal and transverse stiffeners as well as the local compression due to the wheel load. In the numerical example, this constraint was active only for very high girders ($h > 5$ m), therefore, instead of this constraint, we use the prescription of the maximum height

$$h \leqslant h_{\max} \qquad (5.121)$$

with a numerical value of $h_{\max} = 4000$ mm.

Furthermore, the numerical calculations have shown that a deflection constraint with a maximum value of $w_{\max} = L/800$ was not active.

5.5.8 Solution of the Numerical Example

The objective function (5.99) is to be minimized taking into account the constraints (5.105), (5.109), (5.111), (5.117), (5.118) and (5.121). This problem has been solved by backtrack programming with the following series of discrete values (dimensions in mm):

$$h_{\min} = 3200; \qquad h_{\max} = 4000; \qquad \Delta h = 100$$

$$b_{\min} = 200; \qquad b_{\max} = 1000; \qquad \Delta b = 100$$

$$t_{w1}, t_{w2}: 14, \quad 16, \quad 18, \quad 20, \quad 22, \quad 25 \quad 30, \quad 35$$

$$t_{f1}, t_{f2}: \quad 20, \quad 30, \quad 40, \quad 50, \quad 70, \quad 80, \quad 90, \quad 100.$$

The optimum values are as follows: $h = 4000$, $b = 600$, $t_{w1} = 22$, $t_{w2} = 18$, $t_{f1} = 70$, $t_{w2} = 50$. $A_{\min} = 149\,333$ mm^2.

In order to investigate the effect of the special fatigue constraints (5.117) and (5.118), the problem was solved without these constraints as well. The results are as follows: $h = 4000$, $b = 1000$, $t_{w1} = t_{w2} = 20$, $t_{f1} = 35$, $t_{f2} = 25$ mm, $A_{\min} = 140\,000$ mm^2. It can be seen that the special fatigue constraints have a significant effect on the optimum structure. Without these constraints the flange thicknesses may be smaller and it is not necessary to apply a thicker upper part of the web. The girder becomes lighter, but the fatigue constraint (5.117) is not fulfilled ($91 > 75$ MPa) and therefore fatigue cracks may occur.

Chapter 6

Statically Indeterminate
Continuous Girders and Frames

6.1 Literature Survey

Some of the better-known books on *the theory of plasticity of structures* include those of Kaliszky (1975), Massonnet and Save (1977) and Horne (1979). The subject of *plastic analysis of frames* has been treated by Cohn and Maier (1977), Morris and Randall (1975), Uhlmann (1979), and Horne and Morris (1981). Finally, problems related to *the stability of frames* were investigated in several conference papers, e. g. (Proc. Liège 1977), (Proc. Budapest 1977) and recently in a world-wide review of stability of metal structures (Stability 1981——82).

The optimum design of statically indeterminate rod structures under predominantly flexural action occupies an important position in the field of structural optimization, as can be seen from Table 6.1 and Fig. 6.1 (Farkas 1982a).

Frames may be optimized on the basis of either elastic or plastic analyses. In the case of optimum plastic design problems linear programming methods may be used; however, only stress constraints may then be treated. Nonlinear constraints (e. g. displacement, frequency or stability) require the use of nonlinear programming or other special methods (see Chapter 2). The mass (by weight or volume) of the structure is generally adopted as the objective function.

A few remarks on some of the papers appearing in Table 6.1 are relevant. The works of Brozzetti and Lescouarc'h (1973), and Polizzotto and Mazzarella (1979) include the $M-N$ interaction effect on the plastic limit state of rolled I-sections. In the paper of Khalifa and Merwin (1977) the special constraint of torsional buckling of rods is considered in conjunction with an equation expressing that the torque in all members is to be zero. In the article of Miller and Moll (1979) a problem of 14 variables is solved for 6 load conditions; consideration is there given to stress, displacement and local buckling constraints. In most papers the relationships $A-I_x$ and $A-W_x$ of rolled I-sections are treated as continuous and, in some cases, linear functions.

In general, the cross-sectional dimensions differ for each member in the structures considered. In some papers, however, rods of piecewise constant cross-section are allowed for. Venkayya (1971) and Lipp (1976) have treated the frames

Table 6.1

Data for some of the frame structures studied in the literature. *Abbreviations*: st — storey; sp — span; *analysis*: El — elastic; Pl — plastic; *constraints*: S — stress; D — displacement; Fr — frequency; E — earthquake; R — reactions; *math. method*: LP — linear programming; SLP — sequential linear programming; OC — optimality criteria; grad — gradient; SepLP — separable LP; di — with discrete variables; SUMT, Box — see Chapter 2;   anal — analytical method; graph — graphical method

Reference		Structure investigated (Fig. 6.1)	Analysis	Con-straints	Math. method
Brown and Ang	1966	(a), (k)	El	S, D	grad
Bigelow and Gaylord	1967	(d), (f), (k)	Pl	S	LP
Toakley	1968	(g), (h), (j) 4-st	Pl	S	LP di
Kavlie	1970	Ship frames	El	S	SUMT
Venkayya	1971	(k), (m)	El	S, D	OC
Kuzmanovic and Willems	1972	(a), (d), (f), (g), (n), (c) 4-sp	Pl	S	LP
Cho	1972	(c), (g); 2-st 3-sp	Pl + El	S, D	LP di
Majid and Anderson	1972	(b)	El	S, D	SepLP
Venkayya *et al.*	1973	(a)	El	Fr	OC
Brozzetti and Lescouarc'h	1973	(n)	Pl	S	LP
Holst	1974	(g);10-st 3-sp; 30-st 2-sp	El	S, D	SLP
Majid	1974	(b), (d)	El	S, D	graph
Shaw	1974	6-st 3-sp	El	Fr	OC
Shamie and Schmit	1975	(a)	El	S, Fr	SUMT
Frind and Wright	1975	(m), (j) 10-st	El	S, D	grad
Cassis and Schmit	1976b	(a), (g), (h), (j) 7-st	El	S, D, Fr	SUMT
Anraku	1976	(j) 3-st	EL + Pl	E	SLP
Nakamura and Nagase	1976	multi-st, multi-sp	Pl	R	LP
Lipp	1976	(k), (m)	El	S, D	OC
Cheng and Botkin	1976	(h) (sym), (k), (j) 4-st	El	S, D, Fr	grad
Khalifa and Merwin	1977	(p)	Pl	S + spec	LP + grad
Wright and Hakim	1978	(m); (j) 10-st	El	S, D	grad, di
Miller and Moll	1979	(e)	El	S, D	SUMT
Khan *et al.*	1979	(a); frame with 25 rods	El	S, D	OC
Levey and Fu	1979	(a), (g), (h) (sym)	Pl	S	Box, di
Polizzotto and Mazzarella	1979	(j) 3-st	Pl	S	LP
Anderson and Islam	1979	6-st 2-sp	El	D	anal
Saka	1980b	(a), (g), (j) 4-st	El	S, D	LP
Davidson *et al.*	1980	(g)	El	S, D, E	SUMT
Farkas and Szabó	1980	(b) (haunched)	El	S	backtrack
Imai and Shoji	1981	4-st 4-sp (asym)	El	S	dual
Tabak and Wright	1981	(j) 10-st; 15-st 3-sp	El	S, D	OC
Liebman *et al.*	1981	1-st 4-sp; 8-st 3-sp	El	S	SUMT, di

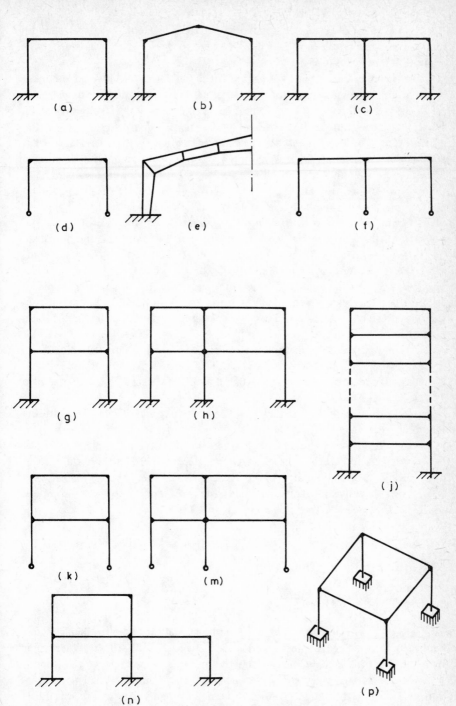

Fig. 6.1 – Some frames studied in the literature

shown in Fig. 6.1k (12 elements) and Fig. 6.1m (18 elements). Clearly, from the point of view of fabrication, it is not advisable to design structures made up of many different types of element.

This aspect is illustrated by the paper of Dupuis (1971), in which the minimum weight design is studied for the case of a uniformly loaded beam constructed from piecewise constant sections. The beam was clamped at one end and simply supported at the other, a deflection constraint at midspan being prescribed; a sandwich cross-section   with variable flange thickness, but constant height and width, was considered. The minimum volumes corresponding to 1, 4, 8, 16 and 32 equal-length parts of different flange thickness were as follows: 1; 0.8866; 0.7959; 0.7696 and 0.7598, respectively. Thus, it can be seen that it is uneconomical to use too many parts of different size in a design.

The basic formulae of *optimality-criteria methods* in the case of frames are the same as those for trusses which were discussed in Chapters 2 and 4. Instead of (4.6) the formula (4.5) may be used with matrices derived for finite elements subjected to bending (see e. g. (Majid 1974)). In Eq. (4.13), valid for a single displacement constraint, I_k should be used instead of A_k. This method was applied to the optimization of a continuous beam with a sandwich cross-section, considering stress constraints and a single displacement limitation, by Dafalias and Dupuis (1972).

The optimality-criteria method was generalized for optimum design of building frames by Tabak and Wright (1981). Continuous beams were optimized both by means of elastic design (Azad 1980) and by plastic design (Fukumoto and Ito 1981).

It is worth noting that the paper of Lee and Knapton (1974) deals with the minimum cost design of a whole structural system of an industrial building using the Box-method. As a result of their study of a particular numerical example, the total cost could be broken up as follows: decking: 80%; main portal frames: 6%; purlins: 5%; production: 7%; wind bracing and other structural steelwork: 2%. Thus it is very important to give also due consideration to the optimization of the decking (see e. g. (Seaburg and Salmon 1971)).

6.2 Optimum Plastic Design

6.2.1 General Aspects

By using the *kinematic theorem* of plastic limit analysis and introducing plastic hinges, the structure is transformed into a chain of rods with one degree of freedom. For this collapse mechanism the external work done by the loads is equal to the internal work at the plastic hinges. Based on this virtual work equation, the ultimate bending moment can be determined which will fix the cross-section to be used in the design. Conversely, on the basis of the *static theorem* a bending moment distribution may be found in equilibrium with the applied loads and satisfying the yield criterion.

In the *minimum volume design* of rod structures with piecewise constant cross-sections the objective function is given by

$$V = \sum_{i=1}^{n} A_i l_i$$

where l_i is the length of a segment and A_i denotes the cross-sectional areas (variables). For a welded I-section optimized for bending on the basis of a stress constraint we have

$$W_{xp} = h^2 t_w / 4 + b t_f h \geqslant W_{op} = M_p / R_y$$

while the web buckling constraint is

$$t_w / h \geqslant \beta_p .$$

In the above expressions M_p is the plastic moment and β_p is the limiting plate slenderness for the plastic range. The minimum cross-sectional area may be expressed as

$$A_{\min p} = \sqrt[3]{13.5 \beta_p W_{op}^2} , \qquad \text{i. e.} \qquad A \sim M_p^{2/3} .$$

so that the objective function becomes

$$F = \sum_{i=1}^{n} M_{pi}^{2/3} l_i .$$

According to Prager (1956) and Massonnet and Save (1977), in the case of rolled I-sections this function can be used in its linearized form, i. e.

$$F = \Sigma M_{pi} l_i .$$

For purposes of optimum plastic design both static and kinematic methods can be used. For larger systems, in which a large number of possible failure mechanisms exists, the static formulation is more suitable.

To begin with, the possible locations of plastic hinges should be determined: these include clamping or supporting points, points at which concentrated forces act, points at which the shear force is zero, etc. At these points (the total number of which will be denoted by m) the bending moment consists of two parts:

$$M_j = \lambda M_{0j} + \sum_{k=1}^{r} a_{jk} X_k .$$

The first term expresses the bending moment due to external proportional loads (λ load factor) while the second denotes the moment due to internal unknown moments (or forces) which number is r. For each M_j, two constraints should be defined:

$$|M_j| \leqslant M_{pj} \begin{cases} +M_j \leqslant M_{pj} \\ \\ -M_j \leqslant M_{pj} \end{cases}$$

Thus, the following linear programming problem should be solved by determining the r unknowns X_k and the m unknowns M_{pj}:

$$\text{minimize} \quad F = \sum_{j=1}^{m} M_{pj} l_j$$

subject to

$$- \sum_{k=1}^{m} a_{jk} X_k - \lambda M_{0j} + M_{pj} \geqslant 0$$

and

$$\sum a_{jk} X_k + \lambda M_{0j} + M_{pj} \geqslant 0$$

$$M_{pj} \geqslant 0.$$

Although continuous beams and frames are mainly loaded in bending, the effects of normal and shear forces may sometimes be significant. Interaction formulae are available as, for example, in (Plastic Design 1971), (Horne 1979), (Lescouarc'h 1977), (Manual 1976). The European Recommendations (1978) and the Hungarian rules for plastic design of steel structures MI−04−188 give such formulae in the following approximate manner:

$$M_u = 0.9 R_u W_p \; ; \qquad N_u = A R_u \; ; \qquad Q_u = \tau_u h t_w = 0.6 R_u h t_w \; .$$

It should be noted that these formulae are taken from the Hungarian rules; however, the Eur.Recom. (1978) are identical provided R_u is replaced by R_y.

(1) *Bending and normal force $(M + N)$*:

$$M_{Nu} = M_u \qquad\qquad \text{for} \quad |N/N_u| \leqslant 0.1$$

$$M_{Nu} = 1.1 (1 - |N/N_u|) M_u \qquad \text{for} \quad |N/N_u| > 0.1 \; .$$

(2) *Bending and shear force $(M + Q)$*:

$$M_{Qu} = M_u \qquad\qquad \text{for} \quad Q/Q_u \leqslant 1/3$$

$$M_{Qu} = M_u (1.1 - 0.3 Q/Q_u) \qquad \text{for} \quad Q/Q_u > 1/3 \; .$$

In the case of $M + N + Q$ the expression for M_{Qu} should be inserted into M_{Nu} instead of that for M_u.

In general, the effect of normal and shear forces on the optimum values is not significant. Therefore, the optimum values corresponding to bending only may simply be corrected. However, if the effect is significant, an iterative procedure should be used.

6.2.2 Example of a Two-Span Box Girder

Figure 6.2 shows a symmetrical two-span girder constructed from two welded box cross-sections. The factored load is $F = 450$ kN and the half span is $l = 6$ m.

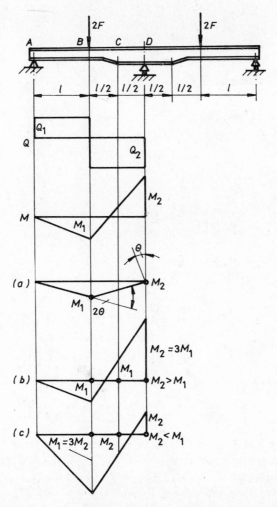

Fig. 6.2 – Symmetrical two-span girder constructed from two box cross-sections; distribution of bending moments and shear forces; (a)–(b)–(c): plastic failure mechanisms

The half volume is taken as the objective function. Assuming that the cross-section of the transitional part is A_1, we obtain:

$$V/l = 1.5A_1 + 0.5A_2 \sim 1.5M_1^{2/3} + 0.5M_2^{2/3}$$

or, approximately,

$$V/l \sim 1.5M_1 + 0.5M_2.$$

Figure 6.2 shows the two plastic failure mechanisms. These are:
(a) Plastic hinges at points B and D. The virtual work equation is written as

$$2F\theta l = 2\theta M_1 + \theta M_2$$

from which

$$M_2 = 2Fl - 2M_1.$$

This equation is illustrated in Fig. 6.3 by the straight line a.

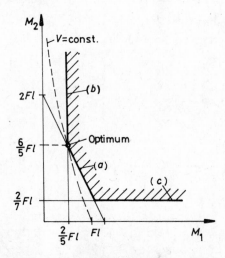

Fig. 6.3 – Optimum plastic design by graphical means in the case of the girder shown in Fig. 6.2

(b) Assuming that $M_2 > M_1$, a plastic hinge can also be formed in the transitional part with plastic moment M_1, so the bending moment diagram shown in Fig. 6.2(b) is valid; then $M_2 = 3M_1$, $M_1 = 2Fl/5$ (line b in Fig. 6.3).
(c) If $M_2 < M_1$, then, according to Fig. 6.2(c), $M_1 = 3M_2$, $M_2 = 2Fl/7$ (line c in Fig. 6.3).
Figure 6.3 shows the feasible region. The line $1.5M_1 + 0.5M_2 =$ const.

touches this feasible region at the optimum point so that

$$M_{1\,opt} = 0.4Fl \;; \qquad M_{2\,opt} = 1.2Fl \;; \qquad M_{1\,opt}/M_{2\,opt} = 1/3 \;.$$

For a box cross-section optimized for bending we have

$$h = \sqrt[3]{W_p/\beta_p} \;; \qquad A = \sqrt[3]{27\beta_p\, W_p^2} \;; \qquad W_p = \gamma_p M/R_y \;; \qquad t_w/2 = \beta_p h \;.$$

Taking $\gamma_p = 1.33$, $R_y = 240$ MPa, $\beta_p = 1/70$; $M_{1\,opt} = 1080$ kNm; $M_{2\,opt} = 3240$ kNm, so that $W_{p1} = 5985$ cm^3, $W_{p2} = 17955$ cm^3; $A_1 = 239.96$ cm^2; $A_2 = 499.13$ cm^2; $h_1 = 74.8$ cm; $h_2 = 107.9$ cm; $t_{w1}/2 = 1.07$; $t_{w2}/2 = 1.54$ cm.

The shear force is $Q = 2F - M_1/l = 720$ kN. At cross-section 1 this gives

$$Q_{u1} = 0.6R_u h_1 t_{w1} = 1920 \text{ kN} \approx 3Q = 2160 \text{ kN} \;,$$

while for section 2 it is $Q_{u2} = 3988$ kN $> 3Q$ so that the M–Q interaction can be neglected.

$$V_{min}/l = 1.5 \times 240 + 0.5 \times 500 = 610 \text{ cm}^2.$$

In the case of a beam with *constant* cross-section we obtain $M_p = Fl/1.5 = 180$ kNm, $W_p = 10^4$ cm^3, $A = 337.9$ cm^2, $V/l = 2A = 675.8$ cm^2. Hence we conclude that the beam with two different cross-sections is 10% lighter than the beam of constant cross-section.

The optimum elastic design of the same girder will be treated in Section 6.3.2.

6.2.3 Example of a Single-Bay Pitched-Roof Portal Frame

The plastic analysis of the portal frame shown in Fig. 6.4 was treated by Morris and Randall (1975). With a snow load of 1 kN/m^2 and with a frame spacing of 4.6 m, the intensity of the factored uniform load is

$$p = (1.1 \times 0.57 + 1.4 \times 1.0)4.6 = 9.3242 \text{ kN/m} \;.$$

The plastic limit load is $p_p = \gamma_p p = 1.3 \times 9.3242 = 12.40$ kN/m. We take the haunch length $c_1 = 2\cos\phi = 1.92$ m, and the distance between the point of maximum bending moment and point D as $c_3 = 2.3$ m.

Calculations have shown that the superposition of the wind load and the temperature effect does not cause bending moments greater than those produced by the vertical load only. Calculations also showed that, for optimum design, it is sufficient to consider only the two symmetrical plastic failure mechanisms illustrated in Figs 6.5 and 6.6.

For the failure mechanism shown in Fig. 6.5 we have

$$c_2 = c_1 \frac{a + d - c_3 \tan\phi}{a + c_1 \tan\phi} = 2.503 \text{ m}$$

and

$$\vartheta = \theta \frac{m}{n} = \theta \frac{a + c_1 \tan\phi}{d - (c_1 + c_3 \tan\phi)} = 3.291\theta \;.$$

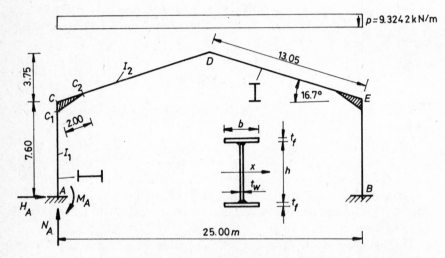

Fig. 6.4 – Single-bay pitched-roof portal frame with fixed bases constructed
from two welded I-sections

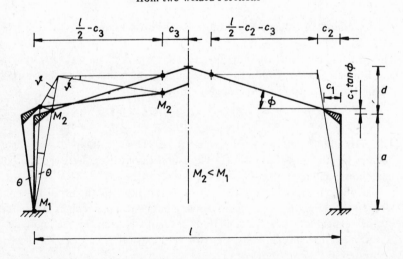

Fig. 6.5 – Symmetrical failure mechanism for the case $M_2 < M_1$

The virtual work equation for half of the structure takes the following form:

$$p_p\left(\frac{l}{2}-c_2-c_3\right)^2\frac{\vartheta}{2}+p_p\left(\frac{l}{2}-c_2-c_3\right)c_3\vartheta-p_p(c_2-c_1)^2\frac{\vartheta}{2}-$$

$$-\frac{p_p}{2}c_1^2\theta=M_1\theta+M_2(\theta+\vartheta)+M_2\vartheta\ .$$

By inserting the above numerical values we get

$$M_2=250.77-0.13189M_1\ .$$

In the case of the failure mechanism shown in Fig. 6.6, and taking a haunch length of $e=1.0$ m, we obtain

$$\theta=\vartheta\,\frac{d+e-c_3\tan\phi}{a-e}=0.61515\,\vartheta\ .$$

The virtual work equation may be written as

$$p_p\left(\frac{l}{2}-c_3\right)^2\frac{\vartheta}{2}+p_p\left(\frac{l}{2}-c_3\right)c_3\vartheta=M_1(\theta+\theta+\vartheta)+M_2\vartheta\ .$$

With the values given above one obtains

$$M_2=935.95-2.2303M_1\ .$$

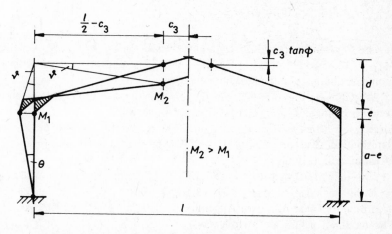

Fig. 6.6 – Symmetrical failure mechanism for the case $M_2>M_1$

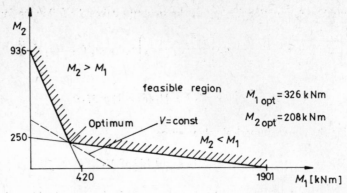

Fig. 6.7 – Optimum plastic design by graphical means of the portal frame shown in Fig. 6.4

The feasible region is given in Fig. 6.7. The optimum plastic moments are: $M_{p1} = 326.3, M_{p2} = 208.1$ kNm.

For purposes of comparison, we calculate the plastic moments using the static method, and consider the equilibrium of the half structure shown in Fig. 6.8. The moment due to the applied load is $p_p l^2/8 = 968.75$ kNm while the moments at points $A - C_1 - C_2 - D_1$ are as follows:

$$A: \quad 968.75 - (M + 11.35 F_A) = -M_{p1}$$

$$C_1: \quad 968.75 - [M + (3.75 + e)F_A] = M_{p1}$$

$$C_2: \quad \frac{p_p}{2}\left(\frac{l}{2} - c_1\right)^2 - [M + (3.75 - c_1 \tan\phi)F_A] = M_{p2}$$

$$D_1: \quad \frac{p_p}{2} c_3^2 - (M + c_3 F_A \tan\phi) = -M_{p2}.$$

The solution of these four equations gives $M = 172.24$ kNm, $F_A = 98.94$ kN, $M_{p1} = 326.54$ kNm, $M_{p2} = 207.72$ kNm. This result agrees well with plastic moments obtained by the kinematic method.

According to Fig. 6.8, the cross-section A of the stanchion is subjected to the bending moment $M_{p1} = 326.54$ kNm, normal force $N_1 = 155$ kN and shear force $Q_1 = 98.94$ kN. The dimensions of the cross-section can be determined by iteration, because the value of $\beta_p = 1/70$ should be modified by considering the effect of compression.

With $\beta_p = 1/70$ we get $W_{p1} = M_{p1}/R_y = 1360$ cm^3, $A_1 = (13.5\beta_p W_{p1}^2)^{1/3} =$ $= 70.92$ cm^2; $N_{u1} = A_1 R_u = 1418$ kN; $N_1/N_{u1} = 155/1418 = 0.11 < 0.27$; according to Table A3.2 (see Appendix A3.2) $1/\beta_{p1} = 70 - 100 N_1/N_{u1} = 59$. The modified values are as follows: $A_1' = 75.08$ cm^2, $h_1 = (2 W_{p1}/\beta_{p1})^{1/3} =$

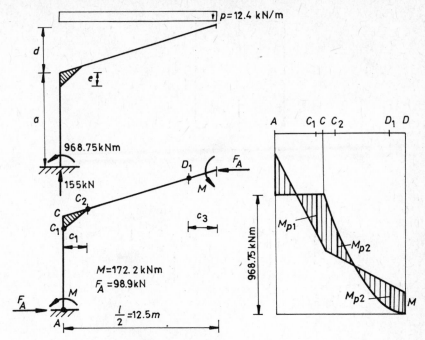

Fig. 6.8 – Minimum volume mechanism for the frame of Fig. 6.4

$= 54.34$ cm, $t_{w1} = \beta_{p1}h_1 = 0.92$ cm; $Q_{u1} = 0.6R_u h_1 t_{w1} = 600$ kN $> 3Q_1$; $N_1/N_{u1} = 0.103 \approx 0.11$, and hence further iteration is not necessary.

At point C_2 the rafter is subjected to $M_{p2} = 207.72$ kNm, $N_2 = 132.17$ kN, $Q_2 = 96.35$ kN. $W_{p2} = 865.5$ cm^3, $A_2 = 52.47$ cm^2, $N_{u2} = 1049$kN, $N_2/N_{u2} = 0.126$; $1/\beta_{p2} = 57$; $A_2' = 56.19$ cm^2, $h_2 = 46.2$ cm, $t_{w2} = 0.81$ cm, $Q_{u2} = 450$ kN $> 3Q_2$; thus, further iteration is not required.

The above values can be rounded off in cm as follows: $h_1 = 55$; $t_{w1} = 0.9$; with $\delta_p = 1/17$; $b_1 = h_1 \sqrt{\beta_{p1}/(4\delta_{p1})} = 14$; $t_{f1} = 0.9$; $h_2 = 47$; $t_{w2} = 0.8$; $b_2 = 11$; $t_{f2} = 0.8$.

The half volume is then $V/2 = aA_1' + s_1 A_2' = 128\,808$ cm^3. The optimum elastic design for this problem will be discussed in Section 6.3.3.

6.2.4 Example of a Closed Frame

In Section 6.3.4 the detailed structural synthesis of closed press frames constructed from two welded box cross-sections (Fig. 6.9(a)) is given. This synthesis is now treated from the viewpoint of plastic design.

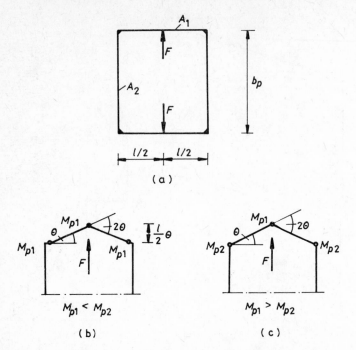

Fig. 6.9 – Failure mechanisms for a doubly symmetric closed frame with members made of two box-sections

If $M_{p1} < M_{p2}$, the equation of virtual work for the failure mechanism shown in Fig. 6.9(b) may be written as

$$F\theta \, l/2 = 4M_{p1}\theta \, ,$$

from which we obtain

$$M_{pl} = Fl/8 \, .$$

This should, however, be modified by considering the effect of   shear.   If one assumes that M_{p1} and $h_1 t_{w1}$ may be expressed in terms of A_1 by using the constants c_1 and c_w, respectively, we obtain $M_{p1} = c_1 A_1$ and $h_1 t_{w1} = c_w A_1$ and the modified equation takes the form

$$\frac{Fl}{8} = M_{p1}\left(1.1 - 0.3 \, \frac{F/2}{0.6R_u h_1 t_{w1}}\right) = c_1 A_1 \left(1.1 - \frac{F}{4c_w R_u A_1}\right) \, ,$$

from which we get

$$A_1 = \frac{F}{1.1 c_1}\left(\frac{l}{8} + \frac{c_1}{4c_w R_u}\right) = A_{10} = \text{const.}$$

If $M_{p1} > M_{p2}$, the kinematic equation of the collapse mechanism shown in Fig. 6.9(c) is given by

$$F\theta l/2 = 2\theta (M_{pl} + M_{p2})$$

i. e.

$$M_{p1} + M_{p2} = Fl/4 \ .$$

Instead of M_{p1} and M_{p2}, the modified formulae for M'_{p1} and M'_{p2} should be used. Inserting M'_{p1} and

$$M'_{p2} = 1.1 M_{p2}\left(1 - \frac{F/2}{A_2 R_u}\right) = 1.1 c_2 A_2\left(1 - \frac{F}{2A_2 R_u}\right)$$

we obtain

$$A_2 = -\frac{c_1}{c_2}A_1 + A_{20} \ ; \qquad A_{20} = \frac{1}{1.1 c_2}\left(\frac{Fl}{4} + \frac{c_1 F}{4 c_w R_u} + \frac{1.1 F c_2}{2 R_u}\right).$$

Figure 6.10 shows the feasible region determined by the straight lines of the above two failure mechanisms and by the line corresponding to the size constraint

$$A_2 \geqslant A_{2\,min} \ .$$

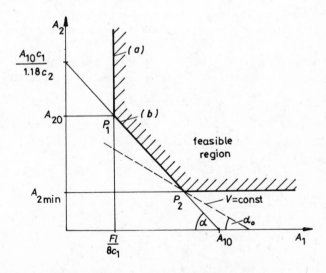

Fig. 6.10 – Optimum plastic design by graphical means of the frame of Fig. 6.9

We take the volume as our objective function so that:

$$\frac{V}{2l} = A_1 + \frac{b_p}{l} A_2.$$

The following two cases are then possible.

(1) If $|\tan\alpha| = c_1/c_2 < |\tan\alpha_0| = l/b_p$, the optimum is determined by the point P_1. Assuming that $c_1 = c_2$ (approximately), the above condition takes the form $b_p < l$. Thus, if $b_p < l, A_{2\,\mathrm{opt}} > A_{1\,\mathrm{opt}}$. This situation is characteristic for portal frames of typical industrial buildings (see Section 6.2.3).

(2) If $|\tan\alpha| = c_1/c_2 > |\tan\alpha_0| = l/b_p$, the optimum is given by the point P_2. Assuming that $c_1 = c_2$, we get $b_p > l$. For this case $A_{2\,\mathrm{opt}} < A_{1\,\mathrm{opt}}$; this situation is characteristic for press frames. Thus, in structural synthesis of press frames with dimensions $b_p > l$, it can be assumed that $A_2 < A_1$, i. e. $I_2 \ll I_1$. With this assumption the optimum elastic design procedure is considerably simplified (see Section 6.3.4).

6.3 Optimum Elastic Design

6.3.1 Iterative Method Using Suboptimized Profiles

The optimum elastic design of a rod structure constructed from n components of welded I- or box profiles of constant cross-section can be performed as a nonlinear programming problem with $4n$ variables. However, by using suboptimized profiles the number of variables may be reduced to n.

In the problem of *suboptimization* one minimizes the cross-sectional area for given $M - N - Q$-values subject to stress and local buckling constraints. The effect of shear forces on web buckling can usually be neglected (see Appendix A3.3).

In this way we have computed a suboptimized series of welded I-sections for the case $Q = 0$ and various M and N values subject to the stress constraint

$$\sigma_N + \sigma_M \leqslant R_u$$

where

$$\sigma_N = N/A ; \qquad \sigma_M = M/W_x ; \qquad A = ht_w + 2bt_f ;$$

$$W_x = h(A_f + A_w/6) ; \qquad A_f = bt_f ; \qquad A_w = ht_w .$$

The web buckling constraint yields (see Appendix A3.2)

$$\frac{h}{t_w} \leqslant 145 \sqrt[4]{\frac{(1 + \sigma_N/\sigma_M)^2}{1 + 173(\sigma_N/\sigma_M)^2}}$$

while the buckling constraint for the compressed flange is simply $b/t_f \leqslant 30$.

Table 6.2

Suboptimized Dimensions of Welded I-sections Subjected to Bending
and Compression for Various Values of the Dimensionless Parameters:
$m = 100\ M/(R_u\ h_0^3) = 5 \times 10^{-4}\ M[kNm]$ and $n = Nh_0/(2M) = 0.5\ N\ [kN]/M\ [kNm]$

m	n	h [cm]	t_w [cm]	$A_f = b\,t_f$ [cm^2]	m	n	h [cm]	t_w [cm]	$A_f = b\,t_f$ [cm^2]
	0.00	56	0.4	7.2		0.00	70	0.5	14.4
	0.05	52	0.4	8.4		0.05	68	0.5	16.0
	0.10	54	0.4	8.0		0.10	64	0.5	18.0
	0.15	50	0.4	9.6		0.15	62	0.5	19.2
0.06	0.20	46	0.4	11.2	0.14	0.20	58	0.5	21.6
	0.25	46	0.4	11.2		0.25	64	0.6	18.0
	0.30	54	0.5	8.0		0.30	72	0.7	14.0
	0.35	48	0.5	10.0		0.35	68	0.7	16.0
	0.40	50	0.5	9.6		0.40	68	0.7	16.8
	0.00	56	0.4	10.8		0.00	72	0.5	16.8
	0.05	56	0.4	11.2		0.05	70	0.5	18.0
	0.10	52	0.4	12.8		0.10	78	0.6	14.0
	0.15	64	0.5	8.0		0.15	74	0.6	16.0
0.08	0.20	60	0.5	9.6	0.16	0.20	68	0.6	19.2
	0.25	46	0.4	16.0		0.25	74	0.7	16.0
	0.30	52	0.5	12.8		0.30	70	0.7	18.0
	0.35	50	0.5	14.0		0.35	68	0.7	19.6
	0.40	50	0.5	14.4		0.40	64	0.7	22.0
	0.00	56	0.4	14.4		0.00	72	0.5	19.2
	0.05	68	0.5	9.6		0.05	68	0.5	22.0
	0.10	54	0.4	16.0		0.10	80	0.6	16.0
	0.15	60	0.5	12.8		0.15	74	0.6	19.2
0.10	0.20	58	0.5	14.0	0.18	0.20	70	0.6	21.6
	0.25	58	0.5	14.4		0.25	74	0.7	19.2
	0.30	54	0.5	16.8		0.30	70	0.7	21.6
	0.35	52	0.5	18.0		0.35	68	0.7	24.0
	0.40	50	0.5	19.2		0.40	66	0.7	25.2
	0.00	72	0.5	11.2		0.00	72	0.5	22.0
	0.05	70	0.5	12.0		0.05	70	0.5	24.0
	0.10	64	0.5	14.4		0.10	88	0.7	14.0
	0.15	62	0.5	16.0		0.15	88	0.7	14.4
0.12	0.20	58	0.5	18.0	0.20	0.20	78	0.7	19.2
	0.25	56	0.5	19.2		0.25	74	0.7	22.0
	0.30	74	0.7	9.6		0.30	72	0.7	24.0
	0.35	64	0.7	14.0		0.35	70	0.7	25.6
	0.40	70	0.7	12.0		0.40	66	0.7	28.8

Table 6.2 (concl.)

m	n	h [cm]	t_w [cm]	$A_f = b\,t_f$ [cm^2]	m	n	h [cm]	t_w [cm]	$A_f = b\,t_f$ [cm^2]
	0.00	84	0.6	21.6		0.00	112	0.8	26.0
	0.05	82	0.6	24.0		0.05	110	0.8	28.0
	0.10	78	0.6	26.4		0.10	112	0.9	26.4
	0.15	86	0.7	21.6		0.15	92	0.8	42.0
0.25	0.20	80	0.7	25.6	0.45	0.20	88	0.8	46.8
	0.25	76	0.7	28.8		0.25	90	0.9	44.8
	0.30	72	0.7	32.0		0.30	84	0.9	51.2
	0.35	74	0.8	30.0		0.35	88	1.0	48.0
	0.40	72	0.8	32.4		0.40	94	1.1	43.2
	0.00	84	0.6	28.0		0.00	100	0.7	39.2
	0.05	84	0.6	28.8		0.05	98	0.7	42.0
	0.10	90	0.7	25.2		0.10	112	0.9	31.2
	0.15	82	0.7	30.8		0.15	94	0.8	46.8
0.30	0.20	88	0.8	26.4	0.50	0.20	96	0.9	44.8
	0.25	82	0.8	31.2		0.25	112	1.1	32.0
	0.30	76	0.8	36.0		0.30	100	1.1	41.6
	0.35	74	0.8	38.4		0.35	100	1.1	44.0
	0.40	78	0.9	35.4		0.40	94	1.1	50.4
	0.00	84	0.6	33.6		0.00	110	0.8	36.0
	0.05	98	0.7	26.4		0.05	110	0.8	38.4
	0.10	90	0.7	31.2		0.10	114	0.9	35.2
	0.15	84	0.7	36.0		0.15	102	0.9	44.8
0.35	0.20	90	0.8	32.0	0.55	0.20	94	0.9	52.8
	0.25	92	0.9	30.8		0.25	108	1.1	40.0
	0.30	86	0.9	35.2		0.30	108	1.2	39.6
	0.35	82	0.9	39.2		0.35	98	1.1	52.0
	0.40	90	1.0	33.6		0.40	94	1.1	57.2
	0.00	110	0.8	22.0		0.00	114	0.8	38.4
	0.05	112	0.8	22.4		0.05	110	0.8	43.2
	0.10	104	0.8	28.0		0.10	108	0.9	44.0
	0.15	92	0.8	36.0		0.15	102	0.9	50.4
0.40	0.20	86	0.8	41.6	0.60	0.20	96	0.9	57.2
	0.25	106	1.1	24.0		0.25	108	1.1	46.8
	0.30	100	1.1	28.8		0.30	110	1,2	44.8
	0.35	100	1.1	30.8		0.35	96	1.1	60.0
	0.40	86	1.0	43.2		0.40	92	1.1	66.0

The backtrack programming method (see Section 2.7) was used for the following list of discrete values of the variables (these are given in cm):

	min	max	Δ
h	40	296	2
b	10	138	2
t_w	0.3	3.5	0.1
t_f	0.4	5.2	0.2

Results appear in Table 6.2.

It is worth noting that Lawo and Thierauf (1978) have published some such data of suboptimized welded I- and box cross-sections subjected to the combined action of bending and normal forces.

If data on suboptimized sections are available, the following iterative procedure can be proposed.

(1) Choose initial values for the ratios of moments of inertia $\omega_i = I_i/I_0$ and determine the appropriate moment, thrust and shear force diagrams. One can either take $\omega_i = $ constant or the ω_i-values obtained for optimum plastic design can be used.

(2) For each component of the structure determine the optimum cross-section for the actual $M - N - Q$ values. Since the optimum dimensions in the table of suboptimized profiles are discrete, it is advisable to use only the optimum dimensions of the web (h, t_w) and to calculate the dimensions of the flanges in the following way. Adopting the notation $\mu_0 = A_f/A_w$, $R_u' = \sqrt{R_u^2 - 3\tau^2}$, $\mu_M = M/(2R_u'A_w h)$, $\mu_N = N/(4R_u'A_w)$, one obtains

$$\mu_0 = \mu_M + \mu_N - \frac{1}{3} + \sqrt{\left(\mu_M + \mu_N - \frac{1}{3}\right)^2 + \mu_M + \frac{\mu_N}{3} - \frac{1}{12}}\,.$$

This calculation results in a fully stressed design.

(3) If the relationships between $A - I_x$ and $A - W_x$ are known, the stress constraints for each member of the structure can be written in terms of ω_i, i. e. with the A_i's as the variables. Then the minimum volume (or cost) design of the whole structure can be carried out by simply considering the stress constraints. For structures constructed from welded I-sections, and subjected predominantly to flexural effects the relationships are as follows (refer to Table 5.1):

$$W_i = (18\beta_i)^{-1/2} A_i^{3/2}\,; \quad I_i = \beta_i A_i^2/3\,; \quad \omega_i = \beta_i A_i^2/(\beta_0 A_0^2)\,.$$

(4) With a new set of ω_i-values calculate again the relevant $M - N - Q$ diagrams and continue this iterative process until the difference between the new and previous A_i-values is less than a prescribed limit.

It should be noted that Step 4 could be left out provided the objective function expresses the volume of the whole structure, since then the fully stressed design agrees, approximately, with the minimum volume design. A numerical example for such a simplified iteration scheme is given in Section 6.3.3.

6.3.2 Example of a Two-Span Box Girder

In Section 6.2.2 the symmetrical two-span girder shown in Fig. 6.11, and constructed from components of welded box cross-section, was optimized on the basis of plastic analysis. It will now be verified that the effect of shear forces on web buckling can be neglected. In the case of both sections we have: $\beta_1 = \beta_2 = 1/145$, $\omega = (A_1/A_2)^2$; $F = 450$ kN; $l = 6$ m. Hence it follows that

$$W_1 = (36\beta)^{-0.5} A_1^{1.5} = 2.0069 A_1^{1.5} ; \qquad W_2 = 2.0069 A_2^{1.5}.$$

In choosing stress constraints one allows for the effect of shearing by specifying a reduced limiting stress value $R'_u = 190$ MPa.

The stress constraint of part 1 of the structure is given by

$$8Fl \frac{4\omega + 1}{37\omega + 27} \leqslant W_1 R'_u$$

from where it follows that

$$A_2 \geqslant A_1 \sqrt{\frac{40000 - 6.5317 A_1^{1.5}}{4.7664 A_1^{1.5} - 10000}}.$$

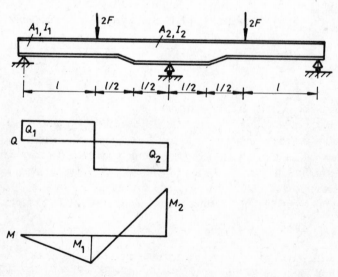

Fig. 6.11 – *M*- and *Q*-diagrams for a two-span girder

For part *2* of the structure one obtains

$$2Fl \frac{5\omega + 19}{37\omega + 27} \leqslant W_2 R_u'$$

so that

$$A_1 \geqslant A_2 \sqrt{\frac{1.9065 A_2^{1.5} - 19\,000}{5\,000 - 2.6127\, A_2^{1.5}}}\,.$$

By using the graphical method, the optimum is found at the intersection of the limiting curves for the two constraints in the coordinate-system $A_1 - A_2$. The result is $A_1 = 232, A_2 = 368$ cm^2; section moduli: $W_1 = 2.0069\, A_1^{1.5} = 7091.8$ cm^3, $W_2 = 14\,168$ cm^3; web dimensions: $h_1 = (0.75\, W_1/\beta)^{1/3} = 91.7 \sim 92$ cm, $t_{w1}/2 = 0.7$ cm, $h_2 = 115.5 \sim 116$ cm, $t_{w2}/2 = 0.8$ cm.

Furthermore, one obtains $\omega = (A_1/A_2)^2 = 0.39745$; $M_1 = 1341.3$ kNm; $M_2 = 2717.4$ kNm;

$$Q_2 = 2F \frac{21\omega + 23}{37\omega + 27} = 676.5 \text{ kN}\,.$$

The dimensions of the flanges can be obtained from the condition

$$\sqrt{\sigma^2 + 3\tau^2} = R_u = 200 \text{ MPa}$$

so that

$$A_f = bt_f = \frac{M}{h\,\sqrt{R_u^2 - 3\tau^2}} - \frac{h t_w}{6}$$

and hence

$$t_f = \sqrt{A_f/30}\,.$$

The results rounded off in cm are: $b_1 = 432; t_{f1} = 14; b_2 = 514; t_{f2} = 18; A_1 = 249.8$ cm^2; $A_2 = 370.6$ cm^2. Finally, the resulting half volume V is

$$V/l = 1.5 A_1 + 0.5 A_2 = 560.0 \text{ cm}^2$$

(c. f. the optimum plastic design which gave 610 cm^2).

Verifications as to the effect of the shear forces can be performed in accordance with the Swiss standard SIA 161 (1979) (see Appendix A3.3).

Cross-section 1. $M_f = R_y bt_f h = 1335$ kNm ≈ 1341.3 kNm. Thus, the effect of shear on web buckling may be neglected provided

$$\tau = \frac{Q}{h t_w} \leqslant \frac{\tau_{cr}}{\gamma_b} = \frac{5.34 \pi^2 E}{12(1 - \nu^2)\gamma_b} \left(\frac{t_w}{2h}\right)^2.$$

With $\gamma_b = 1.08$ we get $52.5 < 54.4$ MPa, so that the effect of shear can be ignored.

Cross-section 2. $M_f = 2576$ kNm < 2717.4 kNm. Therefore, we use the interaction formula (A3.27). $\tau_{cr}/\gamma_b = 44.63$ MPa; $Q_{cr}/\gamma_b = \tau_{cr}ht_w/\gamma_b = 828.5$ kN; $M_w = R_y h^2 t_w/4 = 1292$ kNm; $M_p = M_f + M_w = 3868$ kNm;

$$Q = \frac{Q_{cr}}{\gamma_b} \sqrt{\frac{M_p - M}{M_w}} = 782 \text{ kN} > Q_2 = 676.5 \text{ kN},$$

so that, again, the effect of shear can be neglected. Clearly, no consideration need be given to lateral buckling, due to the large torsional stiffness of the box beam.

6.3.3 Example of a Single-Bay Pitched-Roof Portal Frame

The symmetrical, uniformly loaded, single-bay pitched-roof portal frame, optimized on the basis of plastic analysis in Section 6.2.3, is shown in Fig. 6.12 (see also Fig. 6.4). The frame is constructed from two different welded I-sections. The maximum bending moments in the stanchions and the rafters occur at points A and C_2 respectively. For the elastic analysis of this structure the formulae given in the book by Glushkov *et al.* (1975) can conveniently be used. Introducing the notation $\omega = I_2/I_1$, we obtain

$$|M_A| \text{ [kNm]} = \frac{3266.8895\,\omega + 989.5919}{1.0175\,\omega^2 + 13.86245\,\omega + 0.73031}$$

$$N_A = 116.6 \text{ kN}$$

$$|M_C| \text{ [kNm]} = \frac{4963.8363\,\omega + 88.6588}{1.01750\,\omega^2 + 13.86245\,\omega + 0.73031}$$

$$M_D \text{ [kNm]} = -728.4531 - 0.49342\,|M_A| + 1.49342\,|M_C|$$

$$H_A = (|M_A| + |M_C|)/a$$

$$M_{C2} = |H_A|(a + c_1 \tan\phi) - |M_A| - 116.6c_1 + pc_1^2/2 =$$

$$= 8.17470\,|H_A| - |M_A| - 206.2564$$

$$N_{C2} = (116.6 - pc_1)\sin\phi + |H_A|\cos\phi = 28.3730 + 0.957826\,|H_A|$$

$$Q_{C2} = (116.6 - pc_1)\cos\phi - |H_A|\sin\phi = 94.5347 - 0.287361\,|H_A|.$$

The half volume is used as the objective function, i. e.

$$V/2 = aA_1 + A_2l/(2\cos\phi) = 760A_1 + 1305A_2 \text{ [cm}^3\text{]}.$$

We neglect the effect of shear and consider the following constraints.

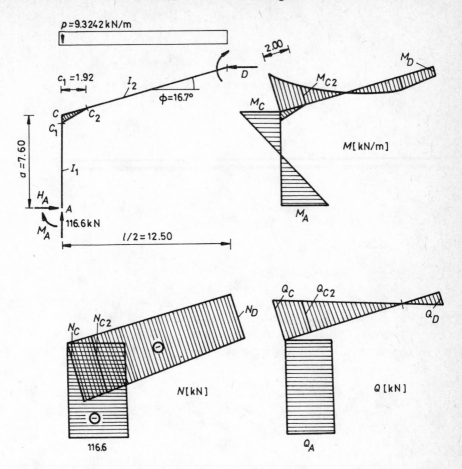

Fig. 6.12 – M–N–Q-diagrams (obtained by elastic analysis) for the portal frame shown in Fig. 6.4

Cross-section 1

   (a) *Stress constraint:*

$$\sigma_{M1} + \sigma_{N1} = M_A/W_1 + N_A/A_1 \leqslant R_u = 200 \text{ MPa}$$

where

$$A_1 = h_1 t_{w1} + 2b_1 t_{f1} ; \qquad W_1 = h_1 (b_1 t_{f1} + h_1 t_{w1}/6) ;$$
$$I_1 = W_1 h_1/2 .$$

   (b) *Web buckling constraint:* according to (A3.14) (see Appendix A3.2) this is

$$\frac{h_1}{t_{w1}} \leqslant 145 \sqrt[4]{\frac{(1 + \sigma_{N1}/\sigma_{M1})^2}{1 + 173(\sigma_{N1}/\sigma_{M1})^2}} .$$

(c) *Flange buckling constraint:* according to (A3.10) we simply have

$$b_1/t_{f1} \leqslant 30 .$$

Cross-section 2
  (a) *Stress constraint:*

$$\sigma_{M2} + \sigma_{N2} = M_{C2}/W_2 + N_{C2}/A_2 \leqslant R_u .$$

Constraints (b) and (c) are the same as those for Section 1, with the appropriate subscript for the variables.

6.3.3.1 Solution by Iteration Using Suboptimized Profiles

Calculations have shown that all constraints are active. So, according to Section 6.3.1, the following iteration procedure may be used.
  (1) Take $\omega = I_2/I_1 = 1$ and determine the $M - N$ diagrams;
  (2) Select suboptimized profiles from Table 6.2 as explained in Section 6.3.1;

Table 6.3

Data on the iteration steps in the optimum elastic design,
with stress constraints, for the portal frame shown in Fig. 6.12

$\omega = I_2/I_1$	Section	M [kNm]	N [kN]	Rounded m	n	h [cm]	t_w [cm]	A_f [cm^2]	I [cm^4]
1	1	273	117	0.14	0.20	58	0.5	21.6	44 461
	2	163	104	0.08	0.35	50	0.5	14.0	22 708
0.5107	1	329	117	0.16	0.20	68	0.6	19.8	61 499
	2	168	111	0.08	0.35	50	0.5	16.0	25 208
0.4099	1	354	117	0.18	0.15	74	0.6	19.2	72 831
	2	167	114	0.08	0.35	50	0.5	16.0	25 208
0.3461	1	375	117	0.18	0.15	74	0.6	21.6	79 402
	2	166	116	0.08	0.35	50	0.5	16.0	25 208
0.3175	1	387	117	0.18	0.15	74	0.6	21.6	79 402
	2	165	117	0.08	0.35	50	0.5	14.4	23 208
0.2923	1	399	117	0.18	0.15	74	0.6	21.6	79 402
	2	164	119	0.08	0.35	50	0.5	14.4	23 208

Table 6.4
Data for the different versions computed by the backtrack method. Sizes in cm

Version No.		h_1	t_{w_1}	b_1	t_{f_1}	h_2	t_{w_2}	b_2	t_{f_2}	Δh Δb	Δt or series	Total number of combinations 10^6	Number of tested combinations 10^3	Run time [s]	$V\min$ 10^3 cm^3
1	min	60	0.4	14	0.4	50	0.4	14	0.4	2 2	0.1	43.0	382.3	1122	134.6
	max	76	1.2	30	1.2	66	1.2	30	1.2						
	opt	64	0.6	24	0.9	50	0.6	14	0.7						
2	min	60	0.4	14	0.4	50	0.4	14	0.4	first 4 4 then 2 2	0.2 0.1	43.0	28.9	124	132.0
	max	76	1.2	30	1.2	66	1.2	30	1.2						
	opt	60	0.5	26	0.9	62	0.6	16	0.6						
3	min	52	0.4	10	0.4	42	0.4	10	0.4	first 8 8 then 2 2	0.4 0.1	1955.1	109.9	398	136.6
	max	84	1.2	42	2.0	74	1.2	42	2.0						
	opt	58	0.5	24	1.1	60	0.5	10	1.6						
4	min	52	0.4	10	0.4	42	0.4	10	0.4	first 8 8 then 2 2	0.4;0.5 0.6;0.8 1.0;1.2 1.4;1.6 1.8	548.0	330.8	1026	133.7
	max	84	1.8	42	1.8	74	1.8	42	1.8						
	opt	58	0.5	22	1.2	52	0.5	12	1.2						
5	min	52	0.5	10	0.5	42	0.5	10	0.5	first 8 8 then 2 2	0.5;0.6 0.8;1.0 1.2;1.4 1.6	200.5	212.0	597	132.8
	max	84	1.6	42	1.6	74	1.6	42	1.6						
	opt	56	0.5	26	1.0	60	0.6	12	0.8						

(3) Calculate a new ω-value and associated $M - N$ diagrams;

(4) Continue with the iteration process until the values of the variables do not change.

The results for the iteration steps are given in Table 6.3. The final results for the variables are (dimensions in cm): $h_1 = 74$; $t_{w1} = 0.6$; $b_1 = 24$; $t_{f1} = 0.9$; $A_1 = 87.6$ cm^2; $h_2 = 50$; $t_{w2} = 0.5$; $b_2 = 18$; $t_{f2} = 0.8$; $A_2 = 53.8$ cm^2.

The half volume is $V/2 = 136\,785$ cm^3 (the optimum plastic design being $V/2 = 128\,808$ cm^3).

6.3.3.2 Solution by Backtrack Programming with 8 Variables

We have computed the above optimization problem in several different ways by using the backtrack programming method (see Section 2.7). Data related to five such modes of computation are given in Table 6.4. The table shows the discrete values, the optimum values, the number of total combinations, the number of combinations actually investigated, the half volume V_{min} and the run time on a CDC 3300 computer.

Comparison of versions 1 and 2 shows that a significant reduction in run time can be achieved by first using a coarse scale of discrete values (larger Δx_i steps) and then continuing with smaller Δx_i values within the smaller region near the optimum determined in the first phase. In versions 3, 4 and 5 the lists of discrete values of plate thicknesses are different. In version 3 the halving method can be directly applied while in version 4 the series is more general although the number of values is now $2^q + 1 = 9$. In version 5 the number of values is only 7 and the series is completed with two 1.6-values.

6.3.4 Example of a Closed Press Frame

6.3.4.1 Design Concepts Relating to Press Frames

Welded frames are widely used for various types of presses of larger load-carrying capacity, e. g. hydraulic presses for vulcanization of rubber, pressing of plastics, etc. Nudel'man (1961) and Nudel'man and Vereshchagin (1963) have worked out a design method for such presses constructed with open-section columns, but Voronin (1966) has pointed out that open-section members are very sensitive to warping torsion resulting from eccentric pressing. Thus, it is more advantageous to use welded box-sections instead of open ones. Kilp's experiments (1970) have also shown that the sensitivity of welded frames due to eccentric pressing is less than that of cast ones.

Here we only consider the structural synthesis of closed frames and hence the method presented is not applicable to the optimum design of open C-frames.

Presses of large load-carrying capacity may consist of two or more welded frames. Some of the features which designers must consider include: a minimum inner space for work (dimensions L and B_p in Fig. 6.13) must be guaranteed; the shear deformations of the cross beams should be taken into account; the techno-logical constraint relating to the ratio of flange thickness to web thickness mentioned in Section 1.2.2 has to be fulfilled.

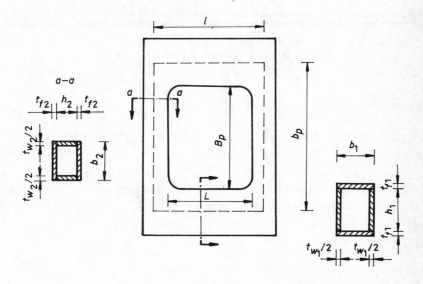

Fig. 6.13 – Main dimensions of a closed press frame constructed from box-sections

In the book by Lanskoy and Banketov (1966) the design of box cross-sections is not treated. The work of Schweer and Mewes (1969) does not take into account shear deformations. Hupfer (1974) and Morgenstern (1975), on the other hand, have considered shear deformations, but their optimized welded I- and box-beams for eccentric presses have been obtained by disregarding local buckling constraints.

Experimental investigations on models have shown that the stresses and deformations can be calculated by treating the structure as a simple closed frame of constant cross-section, provided due consideration is given to the stress concentrations at the corners. These corners should be carefully designed by using rounded, stiffened and thickened inner flanges to avoid larger stress concentrations which could lead to low-cycle fatigue. Some experimental data and associated design concepts may be found in (Puchner and Ruža 1955), (Voronin 1966), (Kilp 1970), (Jesenský 1973), (Geiger 1974).

In the calculations the following assumptions are made.

(1) The frame has three axes of symmetry;

(2) The cross-sections of both transverse girders and columns are constant and doubly symmetric;

(3) The corners are so constructed that a special constraint relating to the stresses at these corners is not required. It is recommended to take a lower value of admissible stress R_{adm} in order to avoid low-cycle fatigue at the corners;

(4) The torsion due to eccentric loading may be neglected;

(5) The load (i. e. pressing force) is uniformly distributed over a considerable part of the cross beams (length $c_L l$, Fig. 6.14). Thus, the maximum bending moment and the maximum shear force do not occur at the same section.

6.3.4.2 Design Constraints and Objective Function

Constraint of maximum bending stress at the centre of the transverse beams:

$$\sigma_{1\,max} = M_1/W_1 \leqslant R_{adm} \qquad (6.1)$$

(subscripts 1 and 2 denote transverse beams and columns respectively). If steel 37 is used it is advisable to take $R_{adm} = 100{-}120$ MPa. The bending moment at

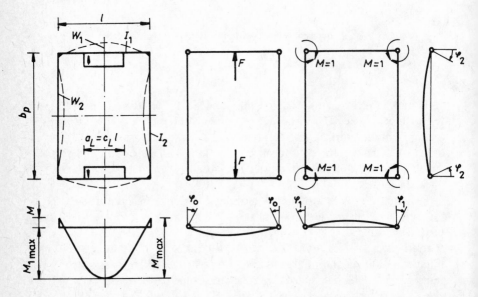

Fig. 6.14 – Determination of corner moments

the corners can be calculated from the following equation relating to the angular deformations shown in Fig. 6.14:

$$M = \frac{\varphi_0}{\varphi_1 + \varphi_2} \tag{6.2}$$

where

$$\varphi_0 = \frac{Fl^2(3 - c_L^2)}{48EI_1} \quad ; \quad c_L = \frac{a}{l} ; \quad \varphi_1 = \frac{l}{2EI_1} ; \quad \varphi_2 = \frac{b_p}{2EI_2}$$

and I_1 and I_2 are the moments of inertia of cross beams and columns respectively. With the definitions $c_b = b_p/l$ and $\vartheta_p = I_1/I_2$, (6.2) becomes

$$M = \frac{Fl}{24} \frac{3 - c_L^2}{1 + c_b \vartheta_p} . \tag{6.3}$$

Thus,

$$M_{1\max} = M_{\max} - M = \frac{Fl(2 - c_L)}{8} - \frac{Fl}{24} \frac{3 - c_L^2}{1 + c_b \vartheta_p} . \tag{6.4}$$

(1) *Constraint of maximum shear stress* in the transverse girders (using an approximate expression for $\tau_{1\max}$):

$$\tau_{1\max} \cong \frac{1.2 Q_{\max}}{A_{w1}} = \frac{1.2F}{2t_{w1}h_1} \leqslant \tau_{\text{adm}} . \tag{6.5}$$

For 37-steels we take $\tau_{\text{adm}} = 92$ MPa.

(2) *Constraint of web shear buckling* (see Appendix A3.3):

$$t_{w1}/(2h_1) \geqslant \beta = 1/90 . \tag{6.6}$$

(3) *Constraint of flange buckling* (see Appendix A3.3):

$$t_{f1}/b_1 \geqslant \delta = 1/30\sqrt{120/200} = 1/40 . \tag{6.7}$$

(4) *Constraint of maximum relative displacement* between the central points of the cross beams:

$$w_1 = w_{1M} + w_{1Q} + w_{2N} \leqslant w_1^* = c_1^* l . \tag{6.8}$$

In this relation c_1^* is the allowable ratio of displacement (refer to Section 1.2.1); in the case of presses for the vulcanization of rubber its value is $1/800 - 1/1000$. The deflection of the cross girder due to bending can be expressed as

$$w_{1M} = \frac{Fl^3(8 - c_L^2 - c_L^3)}{384EI_1} - \frac{Ml^2}{8EI_1} \tag{6.9}$$

where

$$I_1 = \frac{h_1^3 t_{w1}}{12} + 2b_1 t_{f1}\left(\frac{h_1 + t_{f1}}{2}\right)^2. \tag{6.10}$$

According to the results obtained in Section 6.2.4 $I_1 \gg I_2$, so that the second term in (6.4) and (6.9) can be neglected.

The deflection due to shear deformation is given by

$$w_{1Q} = \frac{\rho_q Fl}{8GA_1}(2 - c_L) \tag{6.11}$$

where

$$A_1 = h_1 t_{w1} + 2b_1 t_{f1} \tag{6.12}$$

and

$$\rho_q = \frac{A_1}{I_1^2} \int\limits_{(A_1)} \left(\frac{S_x}{t}\right)^2 dA. \tag{6.13}$$

ρ_q is the coefficient for the shear stress distribution, and S_x is the statical moment. For a box cross-section we obtain

$$\frac{1}{4}\int_A \left(\frac{S_x}{t}\right)^2 dA = \int_0^{b_1/2} \left(\frac{h_1 x}{2}\right)^2 t_{f1}dx + \int_0^{h_1/2}\left[\frac{h_1 b_1 t_{f1}}{2t_{w1}} + \frac{1}{2}\left(\frac{h_1^2}{4} - y^2\right)\right]^2 \frac{t_{w1}}{2}dy.$$

With the definitions $\mu_q = 2t_{f1}/t_{w1}$ and $\zeta_q = b_1/h_1$ ρ_q becomes

$$\rho_q = \frac{3(1 + \mu_q\zeta_q)}{5(1 + 3\mu_q\zeta_q)^2}(2 + 10\mu_q\zeta_q + 15\mu_q^2\zeta_q^2 + 5\mu_q\zeta_q^3). \tag{6.14}$$

Some values of ρ_q appear in Fig. 6.15.

The half-elongation of columns is given by

$$w_{2N} = \frac{Fb_p}{4EA_2} \tag{6.15}$$

where

$$A_2 = h_2 t_{w2} + 2b_2 t_{f2}. \tag{6.16}$$

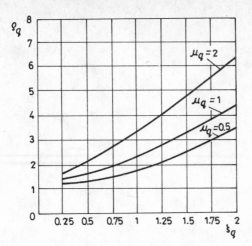

Fig. 6.15 — Diagram for the shear stress distribution factor for a box cross-section

(5) *Constraint of maximum stress in columns:*

$$\sigma_{2\,max} = \frac{F}{2A_2} + \frac{M}{W_{x2}} \leqslant R_{adm} \,.$$ (6.17)

Note that the effect of M in (6.17) should not be neglected.

(6) *Constraint of maximum horizontal displacement* at the centre of a column:

$$w_{2\,max} = \frac{Mb_p^2}{8EI_2} \leqslant c_2^* b_p$$ (6.18)

where $c_2^* = 1/1500 - 1/2000$.

(7) *Fabrication constraints.* In order to simplify the fabrication it is advisable to take

$$b_1 = b_2 = b \,.$$ (6.19)

(8) Since the tendency is to obtain cross-sections for which $I_1 \gg I_2$, a *limit for the ratio* h_2/h_1 should be prescribed so as to ensure a suitable corner joint, i. e.

$$\varphi_h = h_2/h_1 \geqslant \varphi_{h\,min} = 0.3 - 0.5 \,.$$ (6.20)

(9) Furthermore, as pointed out in Section 1.2.2, the following *technological constraints* should be considered:

$$t_{f1} \geqslant 0.7\,t_{w1}/2$$ (6.21)

$$t_{f2} \geqslant 0.7 t_{w2}/2 \,. \tag{6.22}$$

(10) Finally, the *geometrical relations* between the dimension of the inner space of work and that of the frame (Fig. 6.13) are given by

$$b_p = B_p + h_1 \tag{6.23}$$

$$l = L + h_2 \,. \tag{6.24}$$

(11) As *objective function* the volume is selected, i. e.

$$V = 2A_1 l + 2A_2 b_p \,. \tag{6.25}$$

6.3.4.3 Optimum Design by Hand Calculations

Calculations have shown that constraints (6.6), (6.7) and (6.20) can be treated as equalities. Then, considering also (6.19), the number of unknown dimensions can be reduced from eight to four. Furthermore, with the active constraint (6.19), Eq. (6.25) can be written in the form

$$\frac{V}{2l} = A_1 + c_b A_2$$

where

$$c_b = \frac{b_p}{l} = \frac{B_p + h_1}{L + h_1 \varphi_h} \approx \frac{B_p}{L} = \text{constant}.$$

Since A_2 is dependent on formulae (6.37) and (6.38) (to be derived), the constraint

$$A_1 = \text{minimum} \tag{6.26}$$

replaces the more general condition $V = \text{minimum}$. Calculations have also shown that constraints (6.18) and (6.21) are not active. Finally, by neglecting the second term in (6.4) and (6.9) – as mentioned above – the following relatively simple design procedure may be adopted.

(1) The four dimensions of the cross beams can be calculated using (6.1), (6.5), (6.6), (6.7) and (6.26);

(2) With $b_2 = b_1$ and $h_2 = \varphi_{h\min} h_1$, the remaining two unknown dimensions of the column section may be calculated from (6.8), (6.17) and (6.22).

The procedure can be described in more detail as follows.

Treating (6.5) and (6.6) as active, we obtain

$$h_{1Q} = \sqrt{\frac{1.2F}{4\beta \tau_{\text{adm}}}} \,. \tag{6.27}$$

Then, using (6.12), Eq. (6.1) can be written in the form

$$W_{x1} = \frac{2I_1}{h_1} = \frac{h_1^2 t_{w1}}{6} + b_1 t_{f1} h_1 = \frac{A_1 h_1}{2} - \frac{h_1^2 t_{w1}}{3} \geqslant \frac{M_{1\max}}{R_{adm}} . \quad (6.28)$$

From (6.28), and taking into account (6.6), we obtain

$$A_1 = \frac{2M_{1\max}}{R_{adm} h_1} + \frac{4\beta h_1^2}{3} . \quad (6.29)$$

If we now introduce the approximation $M = 0$ in (6.4), and consider relation (6.24), Eq. (6.29) can be written as

$$A_1 = \frac{2C_a}{h_1} + 2C_b + \frac{4\beta h_1^2}{3} \quad (6.30)$$

where

$$C_a = \frac{FL(2 - c_L)}{8R_{adm}} ; \qquad C_b = \frac{F\varphi_{h\min}(2 - c_L)}{8R_{adm}} .$$

From the condition $dA_1/dh_1 = 0$ one obtains

$$h_{1M} = \sqrt[3]{\cdot \frac{3C_a}{4\beta}} = \sqrt[3]{\frac{3FL(2 - c_L)}{32\beta R_{adm}}} . \quad (6.31)$$

The value of h_1 to be adopted should be the *larger* one stemming from (6.27) and (6.31). With this h_1 value expression (6.6) gives

$$t_{w1}/2 = \beta h_1 . \quad (6.32)$$

On the basis of (6.28) and (6.7) we get

$$b_1 = \sqrt{\frac{1}{\delta} \left[\frac{Fl(2 - c_L)}{8R_{adm} h_1} - \frac{h_1 t_{w1}}{6} \right]} \quad (6.33)$$

and

$$t_{f1} = \delta b_1 . \quad (6.34)$$

It can be shown that, in the case of a box-section derived on the basis of an optimum design for bending conditions, the constraint (6.21) is always satisfied. Consider the values $\beta = 1/90$ and $\delta = 1/40$, with which we get $b_1 \geqslant 0.31h_1$. From Table 5.1, $b_1 = h_1\sqrt{\beta/\delta} = 0.67h_1$. Then, if the formula (6.27) gives the larger h_1 value, we obtain, as an approximation, 0.5 instead of 0.67. Thus, the constraint (6.21) is satisfied.

Let us now continue with the computation of the unknowns t_{w2} and t_{f2}. Since

$$W_{x2} \cong \frac{W_{x1}}{\varphi_h \vartheta_p} \quad (6.35)$$

and $\vartheta_p \to \infty$, the constraint (6.17) can be written as

$$\frac{F}{2A_2} + \frac{Fl\varphi_h(3 - c_L^2)}{24W_{x1}c_b} \leqslant R_{adm}.$$ (6.36)

Expression (6.4) with $M = 0$ can be now used so that Eq. (6.28) gives

$$W_{x1} \cong \frac{Fl(2 - c_L)}{8R_{adm}}$$

and, from (6.36), we get

$$A_{2M} = \frac{F}{2R_{adm}\left[1 - \dfrac{\varphi_h(3 - c_L^2)}{3c_b(2 - c_L)}\right]}.$$ (6.37)

On the other hand, from (6.8) and (6.15) one obtains

$$A_{2w} = \frac{Fb_p}{4E(c_1^* - w_{1M} - w_{1Q})}$$ (6.38)

where w_{1M} should be calculated from (6.9) assuming that $M = 0$. Further calculations can then be carried out with the larger of the A_2 values obtained from Eqs (6.37) and (6.38).

From

$$W_{x2} = \frac{W_{x1}}{\varphi_h\vartheta_p} = \frac{A_2h_2}{2} - \frac{h_2^2t_{w2}}{3}$$ (6.39)

we get

$$\frac{t_{w2}}{2} = -\frac{3W_{x1}}{2\varphi_h\vartheta_ph_2^2} + \frac{3A_2}{4h_2}$$ (6.40)

and

$$t_{f2} = \frac{3W_{x1}}{2bh_2\varphi_h\vartheta_p} - \frac{A_2}{4b}.$$ (6.41)

Inserting (6.40) and (6.41) into (6.22) one finally obtains

$$\vartheta_p \leqslant \frac{W_{x1}}{\varphi_hh_2A_2} \cdot \frac{1.050 + 1.500h_2/b}{0.525 + 0.250h_2/b}.$$ (6.42)

Knowing ϑ_p, the unknown dimensions $t_{w2}/2$ and t_{f2} follow from (6.40) and (6.41) respectively.

Numerical example

Data: $F = 5500$ kN; $L = 150$ cm; $B_p = 280$ cm; $R_{adm} = 120$ and $\tau_{adm} = 92$ MPa (i. e. steel 37 is used); $E = 210$, $G = 80$ GPa; $\beta = 1/90$; $\delta = 1/40$; $c_1^* = 1/900$; $c_2^* = 1/2000$; $c_L = 0.3$; $\varphi_{hmin} = 0.4$.

From (6.27) we obtain $h_{1Q} = 127$ cm while, from (6.31), $h_{1M} = 99.5$ cm. Thus, $h_1 = 127$ cm. Condition (6.6) gives $t_{w1}/2 = 1.4$ cm. Then, $h_2 = \varphi_{h\min} h_1 = 50$ cm; $l = L + h_2 = 200$ cm; $b_p = B_p + h_1 = 407$ cm.

According to (6.33), $b_1 = b_2 = b = 62$ cm, and, from (6.7), $t_{f1} = 1.5$ cm. (6.10) gives $I_1 = 1.246 \times 10^6$ cm^4. Assuming that $M = 0$, $w_{1M} = 0.3357$ mm (6.9). With the values $\mu_q = 2t_{f1}/t_{w1} = 1.07$ and $\zeta_q = b_1/h_1 = 0.488$, (6.14) yields $\rho_q = 1.655$.

Relation (6.12) gives $A_1 = 541.6$ cm^2, and, from (6.11), $w_{1Q} = 0.8929$ mm. It can be seen that w_{1Q} is considerably larger than w_{1M}.

$w_1^* = 2.4$ mm; $c_b = b_p/l = 2.0$. On the basis of (6.38), $A_{2w} = 227.2$ cm^2, while, from (6.37), $A_{2M} = 258.7$ cm^2. Thus, $A_2 = 258.7$ cm^2. From (6.42), $\vartheta_p = 11.5$, and hence $\vartheta_p > 10$.

Finally, using (6.40), $t_{w2}/2 = 1.38$ cm (which we round off to 1.4 cm), and, from (6.41), $t_{f2} = 0.97$ cm (rounded off to 1.0 cm).

With these rounded off values, we use (6.10) to get $I_2 = 1.098 \times 10^5$ cm^4, so that $\vartheta_p = I_1/I_2 = 11.35$. Then, from (6.3), $M = 56.28$ kNm, while, from (6.4), $M_{1\max} = M_{\max} - M = 2237.5 - 56.3 = 2181.2$ kNm. It can be seen that the corner moment M is small in comparison with $M_{\max}$.

Check on the fulfilment of constraints

Eq. (6.1): $\sigma_{1\max} = 119 < 120$ MPa.

Eq. (6.8): $w_1 = 0.325 + 0.893 + 1.017 = 2.24$ mm < 2.40 mm.

It can be seen that the effect of shear deformations (second term) and that due to the elongation of the columns (third term) is larger than that corresponding to bending action.

Eq. (6.17): $\sigma_{2\max} = 105.0 + 13.3 = 118.3 < 120$ MPa.

The second term, which expresses the effect of M, is small (but not negligible) as compared to the first term.

Eq. (6.18): $w_2 = 0.505 < 2.035$ mm (not active).

The volume (6.25) calculated with rounded off values is $V = 431\,536$ cm^3.

6.3.4.4 Optimum Design by Backtrack Programming

The problem of 7 variables was also solved by the backtrack method (described in Section 2.7) using the following list of discrete values (dimensions in cm):

	min	max	Δ		min	max	Δ
h_1	60	140	5	b	30	70	5
h_2	30	70	5	t	0.4	1.8	0.2

Optimum dimensions (cm): $h_1 = 115$; $t_{w1}/2 = 1.6$; $b = 60$; $t_{f1} = 1.8$; $h_2 = 50$; $t_{w2}/2 = 1.2$; $t_{f2} = 1.2$; $V = 442\,160$ cm^3.

Chapter 7

Welded Cellular Plates

7.1 Introduction

A cellular plate consists of two parallel face-sheets welded to an orthogonal grid of sheet-ribs sandwiched between them (Fig. 7.1). This type of sandwich plates has the following advantages over plates stiffened on one side: (a) the torsional stiffness is much larger; (b) the height of ribs need be much smaller; (c) because of the symmetry of the structure, initial imperfections due to fabrication (i. e. distortions due to the shrinkage of the welds) are much smaller.

If the height of the ribs is small ($h < 800$ mm), the connection between the face sheets and these ribs can be achieved from the outside by means of arc-spot-welding through the use of a grid of backing strips or by electron-beam welding (see Fig. A5.2 in Appendix A5).

Formulae for the calculation of stresses, deflections, eigenfrequencies and costs in the case of a uniformly loaded, simply supported rectangular cellular plate with different face-thicknesses are summarized in Appendix A5. The author has treated, analytically, the optimum design procedure of such plates (Farkas 1976, 1982b). More recently, detailed analyses were carried out by Shanmugam (1978), Pettersen (1979), and Wilnay (1983).

7.2 Minimum Cost Design of Simply Supported, Uniformly Loaded Square Cellular Plates

7.2.1 Design Constraints and Cost Function

In the case of the symmetrical square cellular plate shown in Fig. 7.1 the four unknowns to be optimized are h, t_f, t_r, a. It is convenient to express these parameters in the following non-dimensional form:

$$\alpha = a/h \ ; \qquad \vartheta = a/t_f \ ; \qquad \mu = t_f/t_r \ ; \qquad \varphi = b/a \ . \tag{7.1}$$

We assume that the upper face plate is subjected to a direct, uniformly distributed normal load. Thus, by using formulae (A5.14), (A5.10), (A5.9) and the values $k_x = 4.79 \times 10^{-2}$, $c_f = 5.13 \times 10^{-2}$ given in Appendix A5, *the constraint of normal stress* in the upper face plate is

$$\sigma_{max} + \sigma_{f\,max} = \frac{4.79 \times 10^{-2}\, p_1\, \alpha\vartheta\varphi^2}{\psi} + 6 \times 5.13 \times 10^{-2}\, \gamma_0\, p_0\, \vartheta^2 \leqslant R_u .$$

(7.2)

σ_{max} arises from the bending of the whole plate, while $\sigma_{f\,max}$ is the normal stress due to the local bending of the upper face sheet. The factored load p_1 can be calculated by means of (A5.16). The effective width ratio ψ may be calculated by formulae (A3.6)–(A3.8) given in Appendix A3.

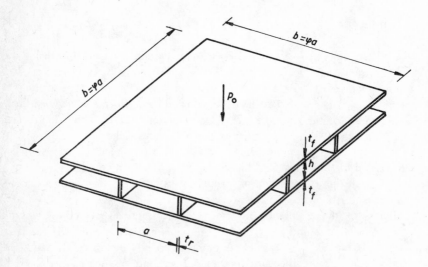

Fig. 7.1 – Welded square cellular plate

On the basis of (A5.13) and (A5.9), and using the value of $k_w = 4.06 \times 10^{-3}$, *the deflection constraint* may be expressed as

$$w_{max} + w_q = 4.06 \times 10^{-3}\, \frac{p_d + p_0}{E_1}\, b\alpha^2\, \vartheta\varphi^3\, \frac{1+\psi}{\psi} + w_q \leqslant c_w^* b , \quad (7.3)$$

where $E_1 = E/(1 - \nu^2)$, c_w^* is the allowable deflection ratio, and $p_d \approx 0.1 p_0$ is the dead weight. w_q is the maximum deflection due to the shear deformation, and is usually negligible. An approximate formula for w_q is given by

$$w_q = \frac{7.37 \times 10^{-2}\, (p_d + p_0)}{G}\, b\alpha\mu\vartheta\varphi . \qquad (7.4)$$

From (A5.18), and with the value of $k_q = 0.42$, *the constraint on the shear buckling of the ribs* can be expressed as

$$\tau = 0.42 p_1 \alpha \mu \vartheta \varphi \leqslant \frac{\tau_{ub}}{\gamma_b} = \frac{5.34 \pi^2 E_1}{12 \gamma_b} \left(\frac{\alpha}{\mu \vartheta} \right)^2 \quad \text{for} \quad \frac{\tau_{ub}}{\gamma_b} \leqslant \tau_u \quad (7.5a)$$

$$\tau = 0.42 p_1 \alpha \mu \vartheta \varphi \leqslant \tau_u \quad \text{for} \quad \frac{\tau_{ub}}{\gamma_b} > \tau_u . \quad (7.5b)$$

By using formulae (A5.21) and (A5.22), *the constraints on the first eigenfrequency* of the whole plate, and of a plate element of the upper face-sheet (assumed to be simply supported), are as follows:

$$f_{11} = \frac{\omega_{11}}{2\pi} = \frac{\pi}{2b} \sqrt{\frac{E_1}{\rho}} \; \frac{1}{\alpha \varphi \sqrt{1 + 1/(\alpha \mu)}} \geqslant f_0 \quad (7.6)$$

and

$$f_{11f} = \frac{\pi}{2b} \sqrt{\frac{E_1}{3\rho}} \; \frac{\varphi}{\vartheta} \geqslant f_0 . \quad (7.7)$$

Finally, *the limitations imposed on the plate thicknesses* are given by

$$t_f \geqslant t_0 \quad \text{or} \quad \vartheta \varphi \leqslant b/t_0 \quad (7.8)$$

$$t_r \geqslant t_0 \quad \text{or} \quad \mu \vartheta \varphi \leqslant b/t_0 . \quad (7.9)$$

The formulae for *the cost function* ((A5.23) and (A5.24)) may be simplified as follows:

$$K = K_1 + K_2 = 2k_1 \rho b^3 \left(\frac{1}{\varphi \vartheta} + \frac{1}{\alpha \mu \vartheta \varphi} \right) +$$

$$+ \frac{4k_2 b \varphi}{\alpha} + 4(k_3 + k_4) b \varphi . \quad (7.10)$$

The cost factors $k_1, ..., k_4$ are defined in Appendix A5.

7.2.2 Numerical Example

Data: $b = 8$ m; $R_u = 200$ MPa, $\tau_u = 115$ MPa, $R_y = 240$ MPa

(i. e. steel 37 is to be used); $E = 210$ GPa,

$E_1 = E/(1 - \nu^2) = 230.77$ GPa; $G = 80$ GPa; $t_0 = 2$ mm;

$c_w^* = 1/200$; $p_0 = 5000$ N/m^2; $p_d = 0.1 p_0$; $\gamma_d = 1.1$;

$\gamma_0 = 1.2$; $\gamma_b = 1.1$; $p_1 = 6550$ N/m^2; $\rho = 7850$ kg/m^3;

$f_0 = 15$ Hz; $\;k_1 = 0.3924$ \$/kg; $\;k_2 = 1.0$ \$/m; $\;k_3 + k_4 = 2.5$ \$/m;

$\eta = 3$ (used in formula (A3.7)).

Note that the welding costs may be considerably higher, if one calculates the whole shop labour cost (see the remark in Section 5.2.8.4).

7.2.2.1 Analytical Solution

We start with the objective function (7.10), which may be written as

$$\frac{K}{512} \; [\$/m^3] = 6160 \left(\frac{1}{\vartheta \varphi} + \frac{1}{\alpha \mu \vartheta \varphi} \right) + 0.0625 \frac{\varphi}{\alpha} + 0.15625 \varphi . \quad (7.11)$$

The stress constraint (7.2) is given by

$$\sigma_{max} + 0.18468 \times 10^{-2} \; \vartheta^2 \leqslant R_u \quad [\text{MPa}] \quad (7.12)$$

where

$$\sigma_{max} = 3.13745 \times 10^{-4} \; \alpha \vartheta \varphi^2 / \psi \quad [\text{MPa}]$$

and

$$\psi = \frac{2}{\lambda p} - \frac{1}{\lambda_p^2} - \frac{6}{\vartheta - 6} \quad \text{for} \quad \lambda_p = \vartheta \sqrt{\frac{\sigma_{max}}{E}} \geqslant \lambda_{p0} = 1.9 \sqrt{\frac{\sigma_{max}}{rR_y}} . \quad (7.13)$$

Since ψ depends on σ_{max}, it is necessary to use an iterative procedure. It is advisable to choose a value for σ_{max}, solve the optimization problem and, using the resulting optimum values, re-calculate the value of σ_{max}. (Note that the iteration may be eliminated by solving Eqs (7.12) and (7.13) for ψ, see Farkas (1982b)).

Knowing σ_{max}, we obtain (from (7.12))

$$\vartheta \leqslant \sqrt{\frac{R_u - \sigma_{max} \; [\text{MPa}]}{0.18468 \times 10^{-2}}} . \quad (7.14)$$

Since the maximum K may be calculated with the value ϑ_{max}, Eq. (7.14) can be treated as an equality. With ϑ known, the ψ value can be obtained, so that Eq. (7.12) takes the form

$$\alpha \varphi^2 \leqslant C_f ; \quad C_f = \frac{10^4 \; \sigma_{max} \; [\text{MPa}] \; \psi}{3.13745 \; \vartheta} = 3187.302 \frac{\sigma_{max} \; [\text{MPa}] \; \psi}{\vartheta} . \quad (7.15)$$

Now, K_{min} may be calculated by using μ_{max}, and hence Eq. (7.9) can be treated as active:

$$\mu = \frac{b}{t_0 \, \vartheta \varphi} = \frac{4000}{\vartheta \varphi} . \tag{7.16}$$

With the introduction of (7.16) into (7.11) the objective function becomes

$$\frac{K}{512} \, [\$/m^3] = \frac{6160}{\vartheta \varphi} + \frac{1.54}{\alpha} + 0.0625 \, \frac{\varphi}{\alpha} + 0.15625 \varphi . \tag{7.20}$$

Furthermore, if one neglects w_q, the deflection constraint (7.3) may be written as

$$\alpha^2 \, \varphi^3 \leqslant C_e ; \quad C_e = \frac{1000 E_1 c_w^* \, \psi}{4.06 (p_d + p_0) \, \vartheta (1 + \psi)} = \frac{5.16724 \times 10^7 \, \psi}{\vartheta (1 + \psi)} . \tag{7.21}$$

When (7.16) is used the rib buckling constraints (7.5a–b) take the form

$$\alpha \varphi^2 \geqslant C_h ; \quad C_h = 191.08 \tag{7.22a}$$

and

$$\alpha \leqslant 10.45 . \tag{7.22b}$$

Comparing (7.15) with (7.22a), we can see that, disregarding the other constraints, the feasible region is defined by parallel curves characterized by C_f and C_h, respectively. It can readily be shown that the approximate objective function

$$K \sim \frac{6160}{\vartheta \varphi} + 0.15625 \varphi$$

becomes a minimum when $C_f = C_h$, i.e. when $\Delta C = C_f - C_h = 0$. According to (7.15) $C_f = c_{f0}/\vartheta$, so that

$$\frac{1}{\vartheta} = \frac{\Delta C + C_h}{c_{f0}}$$

i.e.

$$K \sim \frac{6160}{\varphi} \cdot \frac{\Delta C + C_h}{c_{f0}} + 0.15625 \, \varphi .$$

Thus, $\Delta C = 0$ minimizes K. From this it follows that, in the iterative procedure, σ_{max} or ϑ values should be determined by imposing the condition that $C_f = C_h$. It should be noted that such condition is valid only when the approximate expression for K is adopted.

In the present numerical example we take

$$\sigma_{max} = 50 \text{ MPa}$$

11*

and, from (7.14), $\vartheta = 285$. With $r = 0.6$, (7.13) takes the form

$$\lambda_p = 4.398 > \lambda_{p0} = 1.12$$

and $\psi = 0.38155$. From (7.15) we obtain $C_f = 213.35$. Since $C_f \approx C_h$, further iteration is not necessary.

The objective function (7.20) is given by

$$\frac{K}{512}\ [\$/m^3] = \frac{21.614}{\varphi} + \frac{1.54}{\alpha} + 0.0625\ \frac{\varphi}{\alpha} + 0.15625\,\varphi\ . \qquad (7.23)$$

Stress constraint (7.15): $\qquad\qquad\qquad \alpha\varphi^2 \leqslant 213.35$

Rib buckling constraints (7.22a–b): $\qquad \alpha\varphi^2 \geqslant 191.08$

$$\alpha \leqslant 10.45$$

Deflection constraint (7.21): $\qquad\qquad \alpha^2\,\varphi^3 \leqslant 50072$

Frequency constraints (7.6) and (7.7): $\qquad \alpha\varphi\sqrt{1 + \dfrac{\varphi}{14.035\alpha}} \leqslant 70.977$

and $\qquad\qquad\qquad\qquad\qquad\qquad \varphi \geqslant 6.95\ .$

By plotting the limiting curves for the above constraints in the $\alpha - \vartheta$ coordinate-system, we can conclude that the stress constraint is active. Thus, inserting $\alpha = 213.35/\varphi^2$ into (7.23) we get

$$\frac{K}{512}\ [\$/m^3] = \frac{21.614}{\varphi} + 7.218 \times 10^{-3}\ \varphi^2 + 2.9294 \times 10^{-4}\ \varphi^3 + 0.15625\ \varphi\ .$$

$$(7.24)$$

The condition $dK/d\varphi = 0$ gives $\varphi_{opt} = 8.1$.

The remaining relevant values are: $a = b/\varphi = 98.77$ cm; $t_f = a/\vartheta = 0.347$ cm; $\mu = 14.035/\varphi = 1.735$; $\qquad t_r = t_f/\mu = 0.2$ cm; $\qquad \alpha = 213.35/\varphi^2 = 3.2518$; $h = a/\alpha = 30.37$ cm. The corresponding minimum cost is $K_{min} = 2236.4$ \$.

It can be shown that the cellular plate with dimensions (in mm) $a = 1000$, $t_f = 3.5$, $t_r = 2$ and $h = 310$ fulfils all constraints, while, at the same time, Eqs (7.12) and (7.13) give, approximately, the same value for σ_{max}.

7.2.2.2 Solution by the SUMT Method

The scaled unknowns are as follows: $x_1 = \alpha$; $x_2 = \mu$; $x_3 = \vartheta/100$, $x_4 = \varphi$. The cost function is given by

$$\frac{K}{512} = 61.6\left(\frac{1}{x_3\,x_4} + \frac{1}{x_1\,x_2\,x_3\,x_4}\right) + 0.0625\ \frac{x_4}{x_1} + 0.15625\,x_4\ . \quad (7.25)$$

Inserting the σ_{max} given by (7.12) into (7.13) we obtain a second-order equation for ψ. The approximate solution of this equation is

$$\psi \approx \frac{4Y}{(Y+1)^2}\left[1 - \frac{Z}{2}(Y+1)\right]; \qquad Z = \frac{6}{100x_3 - 6}$$

$$Y = \frac{4.79 \times 10^{-2}\, p_1\, \alpha\varphi^2\, \vartheta^3}{E} = 1.494 \times 10^{-3}\, x_1\, x_3^3\, x_4^2 .$$

Hence the appropriate stress constraint may be written as

$$0.18468x_3^2 + \frac{0.052501(1.494 \times 10^{-3}\, x_1\, x_3^3\, x_4^2 + 1)^2}{x_3^2\left[1 - \dfrac{6}{100x_3 - 6}(1.494 \times 10^{-3}\, x_1\, x_3^3\, x_4^2 + 1)\right]} \leqslant 2 . \quad (7.26)$$

The deflection and frequency constraints were not considered because they were passive in the previous analytical approach to this optimization problem. The rib buckling constraint is given by

$$\frac{x_2^3\, x_3^3\, x_4}{x_1} \leqslant 320.833 , \qquad\qquad (7.27)$$

while the limitations on plate thicknesses take the form

$$x_3\, x_4 \leqslant 40 \qquad\qquad (7.28)$$

and

$$x_2\, x_3\, x_4 \leqslant 40 . \qquad\qquad (7.29)$$

The computed optimum values are: $x_1 = 3.598$; $x_2 = 1.775$; $x_3 = 2.732$; $x_4 = 8.248$. The corresponding minimum cost is $K_{min} = 2351.9$ \$. It can be seen that the results agree reasonably well with the values obtained by the analytical method.

For purposes of comparison, the problem was solved by the SUMT method using *only the material cost* of the plate as the objective function, so that the latter was given by

$$\frac{K_1}{512} = 61.6\left(\frac{1}{x_3\, x_4} + \frac{1}{x_1\, x_2\, x_3\, x_4}\right). \qquad (7.30)$$

The results are: $x_1 = 2.529$; $x_2 = 1.285$; $x_3 = 2.063$; $x_4 = 15.10$. The corresponding value of the total cost is $K = 2723.1$ \$, i. e. 15.8% greater than the above value K_{min}. It can be concluded that a considerable cost saving may be achieved by including the welding costs in the objective function.

Sensitivity analysis. In order to compare the sensitivity of the objective functions K and K_1, we calculate the changes in these functions due to variations in the main parameter $\varphi = x_4$ in the vicinity of φ_{opt}. If $\varphi_{opt} = 8.1$ is changed to $\varphi = 10$ (23%), the value of K in (7.24) increases by 3.8%. An approximate one-variable form of (7.30) may be written as

$$\frac{K_1}{512} = \frac{25}{\varphi} + 3.6 \times 10^{-3} \, \varphi^2 \, .$$

If, in the latter, $\varphi_{opt} = 15.1$ changes by 23%, K_1 increases by 4.5%. Thus, the more complex the objective function, the smaller its sensitivity.

Appendix

A1 Approximate Calculation of Residual Welding Stresses

The method of Okerblom (Okerblom *et al.* 1963), (Farkas 1974a), adequate for purposes of engineering design, is here summarized, being then applied to the calculation of residual compressive stresses in plate elements components of struts and/or beams. It should be noted that here consideration is only given to residual stresses due to longitudinal continuous single-pass welds.

Figure A1.1 shows an I-section beam with a longitudinal weld at the fibre y_A. The residual strain distribution can be seen in Fig. A1.1 (a). These deforma-

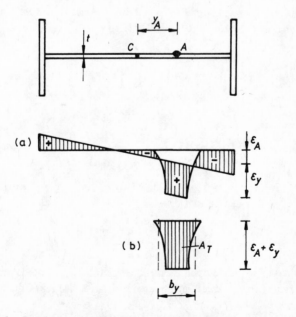

Fig. A1.1 – Residual strain distribution caused by a longitudinal weld in an I-beam

tions are ·caused by the shrinkage force proportional to the area A_T shown in Fig. A1.1 (b). ϵ_A is the strain due to the elastic deformation of the whole beam while ϵ_y is the yield strain. According to Okerblom

$$A_T t = \frac{0.4840\,\alpha_0\,Q_T}{c_0\,\rho}\ln 2 = \frac{0.3355\,\alpha_0\,Q_T}{c_0\,\rho} \tag{A1.1}$$

$$Q_T = \eta_0\,UI/v = q_0\,A_v \tag{A1.2}$$

where α_0 = coefficient of thermal expansion; c_0 = specific heat; ρ = density; U = arc voltage; I = arc current; η_0 = coefficient of efficiency; v = speed of welding; A_v = cross-sectional area of the weld; q_0 = coefficient; t = thickness of the plate.

For mild and low alloy steels $\alpha_0 = 12 \times 10^{-6}/^{\circ}C$; $c_0\,\rho = 4.77 \times 10^{-3}$ J/mm^3/$^{\circ}$C; and,

$$A_T t\,[\mathrm{mm}^2] = 0.8440 \times 10^{-3}\,Q_T \qquad [\mathrm{J/mm}]\,. \tag{A1.3}$$

The elastic deformation of the whole beam due to the eccentric shrinkage effect $A_T t$ can be characterized by the strain at the centroid ϵ_C and the curvature C. According to the theory of thermoelasticity these are given by

$$\epsilon_C = \frac{A_T t}{A} \qquad \text{and} \qquad C = \frac{A_T t y_A}{I_x}\,. \tag{A1.4}$$

A and I_x are the cross-sectional area and the moment of inertia of the beam respectively. The elastic strain at the weld is then

$$\epsilon_A = \epsilon_C + C y_A\,. \tag{A1.5}$$

The average width of the plastic tension zone around the weld (Fig. A1.1 (b)) is given by

$$b_y = A_T(\epsilon_A + \epsilon_y) \tag{A1.6}$$

so that the area of the plastic zone becomes

$$A_y = b_y\,t = A_T t/(\epsilon_A + \epsilon_y)\,. \tag{A1.7}$$

Figure A1.2 shows the simplified residual stress distribution. By using (A1.4) and (A1.5), Eq. (A1.7) can be written as

$$\frac{1}{A_y} = \frac{1}{A} + \frac{y_A^2}{I_x} + \frac{\epsilon_y}{A_T t}\,. \tag{A1.8}$$

If no crookedness is developed in the beam during welding, as, for example, in the case of a symmetrical weld arrangement, Eq. (A1.8) takes the form

$$A_y^{-1} = A^{-1} + \epsilon_y/(A_T t)\,. \tag{A1.9}$$

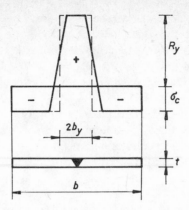

Fig. A1.2 – Simplified residual stress distribution for a central weld

When the beam is infinitely stiff we obtain

$$A_y = A_T\, t/\epsilon_y \,. \tag{A1.10}$$

From the equilibrium equation

$$(b - b_y)\sigma_c = b_y R_y \,,$$

coupled with Eq. (A1.9), one obtains the following expression for the residual compressive stress:

$$\sigma_c = \frac{A_T\, t R_y}{A\,\epsilon_y} = \frac{A_T\, t}{A}\, E = \frac{0.3355\,\alpha_0\,\eta_0\, UIE}{c_0\,\rho v b t} \,. \tag{A1.11}$$

White (1977a) has proposed the following formula:

$$\sigma_c = 0.2\,\eta_0 UI/(v b t) \tag{A1.12}$$

which, with the values $\alpha_0 = 11 \times 10^{-6}$; $E = 2.05 \times 10^5$ MPa; $c_0\,\rho = 3.53 \times 10^{-3}$ J/mm^3/°C (also proposed by White) reduces Eq. (A1.11) to

$$\sigma_c = 0.214\,\eta_0\, UI/(v b t) \,.$$

Therefore, it can be seen that Okerblom's formula is in agreement with the new empirical values.

The formulae used above are valid only if $Q_T/A \leqslant 2.50$ J/mm^3 for symmetrically arranged welds $(y_A = 0)$, and if $Q_T/A \leqslant 0.63$ J/mm^3 for eccentric welds $(y_A \neq 0)$.

In the case of approximate calculations one can use the simple formula

$$\epsilon_y = R_y/E$$

where R_y is the yield stress of the parent plate. The measurements have shown that the *weld metal* may have a yield stress different from that of the parent material. This discrepancy arises due to the difference between the yield stress of the parent material and that of the electrode. For instance, mild steels with a yield stress of 240 MPa welded with electrodes of tensile strength 500–550 MPa may have welds of yield stress of about $R_{y1} = 300$ MPa. Therefore, we propose the use of $\epsilon_{y1} = R_{y1}/E$ in Eqs (A1.8)–(A1.10) instead of ϵ_y, if the R_{y1} value is known from measurements.

In the case of high-strength steels the yield stress of the weld metal is smaller than that of the parent material. Therefore, the residual welding stresses are relatively smaller than those in the case of mild steel.

For automatically welded fillet welds Okerblom has proposed the formula

$$Q_T \text{ [J/mm]} = 59.5 \, a_w^2 \text{ [mm}^2\text{]} \tag{A1.13}$$

where a_w is the minimum dimension of the weld (Fig. A1.3).

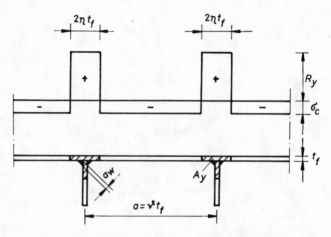

Fig. A1.3 – Residual stress distribution in a stiffened plate

In the case of stiffened plates, and neglecting the cross-sectional area of the ribs (Fig. A1.3), the area of a section of width a is

$$A \cong a t_f = \vartheta \, t_f^2 \,. \tag{A1.14}$$

For double fillet welds, (A1.13) can be used with a correction factor of 1.2. With $\epsilon_y = 1.2 \times 10^{-3}$, (A1.9) becomes

$$\frac{A_T \, t_f}{\epsilon_y} = 1.2 \times 41.7 \, a_w^2 = 50.1 \, a_w^2 \,. \tag{A1.15}$$

The area of the plastic zone (see Fig. A1.3) can be expressed as

$$A_y = \eta(2 \, t_f^2 + t_r^2) \,. \tag{A1.16}$$

By combining (A1.9), (A1.14) and (A1.15) one obtains

$$\eta = \frac{A_y}{2t_f^2 + t_r^2} = \frac{1}{\left[2 + \left(\dfrac{t_r}{t_f}\right)^2\right]\left[\dfrac{1}{\vartheta} + \dfrac{1}{50.1}\left(\dfrac{t_f}{a_w}\right)^2\right]} . \tag{A1.17}$$

For ordinary civil engineering structures typical values of the above parameters are $a_w = 0.45\ t_f$, $\vartheta = 100$, $t_r = t_f$. With these values Eq. (A1.17) gives $\eta = 3.07$. In the case of ship structures one could adopt $a_w = 0.6\ t_f$, $\vartheta = 50$, $t_r = t_f$ so that $\eta = 4.42$. Faulkner *et al.* (1973) have proposed that for lightly welded structures the value $\eta = 3$ be taken, while for heavily welded ones $\eta = 4.5$ is adequate. These values are in good agreement with those obtained from Okerblom's formulae.

A2 Buckling of Compressed Struts

Based on a large amount of experimental work, the European Convention for Constructional Steelwork (ECCS – in French CECM) has proposed three (a, b, c) buckling curves for different types of cross-sections taking into account the effects of initial imperfections and residual stresses (Beer and Schulz 1970), (Proc. Int. Coll. 1975), (Manual 1976).

The buckling criterion is

$$N/A \leqslant \varphi_b\, R_u \tag{A2.1}$$

where N is the factored compressive force, A is the cross-sectional area, while R_u represents the ultimate stress.

$$1/\varphi_b = 0.5 + \alpha_1\,\bar{\lambda}^2 + \sqrt{(0.5 + \alpha_1\,\bar{\lambda}^2)^2 - \beta_1\,\bar{\lambda}^2} \tag{A2.2}$$

where

$$\bar{\lambda}^2 = \frac{R_y}{\pi^2\,E}\,\lambda^2 \tag{A2.3}$$

λ being the column slenderness ratio. The values of α_1 and β_1 are given in Table A2.1.

Dwight (1975), as well as Maquoi and Rondal (1978), have proposed another calculation method based on the equation

$$\frac{N}{A} + \frac{Na_0}{(1 - N/N_E)W} \leqslant R_y , \tag{A2.4}$$

where

$$N_E = \pi^2\,EA/\lambda^2 . \tag{A2.5}$$

Table A2.1
Constants in the Formulae for the European Buckling Curves

Author	Curve	a_0	a	b	c	d
CECM	α_1	–	0.514	0.554	0.532	–
	β_1		0.795	0.738	0.377	
MSZ 15024	α_1	–	0.60	0.65	0.65	–
	β_1		0.80	0.80	0.40	
Maquoi and Rondal (1978)	α_M	0.093	0.158	0.281	0.384	0.587
Braham *et al.*, (1980)	α_B	0.125	0.206	0.339	0.489	0.756

N_E is the Euler buckling load, while a_0 represents the amplitude of a sinusoidal imperfection in the case of a strut pinned at both ends. (A2.4) can be written as

$$(R_y - \sigma)(\sigma_E - \sigma) = \eta_b\, \sigma\, \sigma_E \qquad (A2.6)$$

where the following notation has been adopted:
$\sigma = N/A$, $\sigma_E = N_E/A$, $\eta_b = a_0 A/W$. The solution of (A2.6) gives

$$\overline{N} = \frac{\sigma}{R_y} = \frac{1}{2\,\overline{\lambda}^2}\left[1 + \eta_b + \overline{\lambda}^2 - \sqrt{(1 + \eta_b + \overline{\lambda}^2)^2 - 4\,\overline{\lambda}^2}\right], \quad (A2.7)$$

but $\overline{N} = 1$ $(\eta_b = 0)$ if $\overline{\lambda} \leqslant 0.2$.

According to Maquoi and Rondal: $\eta_b = \eta_M = \alpha_M\sqrt{\overline{\lambda}^2 - 0.04}$, while Braham *et al.* (1980) propose $\eta_b = \eta_B = \alpha_B(\overline{\lambda} - 0.2)$. Values of α_M and α_B are given in Table A2.1.

In members made of *high-strength steels* the residual welding stresses are relatively smaller (see Appendix A1); for this reason another buckling curve (a_0) was given. A fifth buckling curve (d) is also needed for thick-walled members in which the effect of residual welding stresses is larger.

It should be noted that *thin-walled struts of open cross-section* may also fail due to flexural-torsional buckling (Manual 1976).

Another problem is *the interaction* effect of overall and local buckling. If the plate elements of a strut buckle locally during the general buckling process, the failure load may be significantly smaller. On the basis of experiments Braham *et al.* (1980) have proposed a calculation method for cold-formed rectangular hollow sections.

The local buckling strength is given by

$$\overline{N}_v = \frac{\sigma_{krv}}{R_y} = \frac{1}{2\overline{\lambda}_v} \left[1 + \eta_v + \overline{\lambda}_v - \sqrt{(1 + \eta_v + \overline{\lambda}_v)^2 - 4\overline{\lambda}_v} \right] \quad \text{(A2.8)}$$

where

$$\overline{\lambda}_v = \frac{b_{eq}}{1.9t} \sqrt{\frac{R_y}{E}} \; ; \quad \eta_v = \alpha_v(\overline{\lambda}_v - \overline{\lambda}_{vo})$$

$\overline{\lambda}_{vo} = 0.8$; $\alpha_v = 0.67$ and 0.35 for untempered and tempered sections, respectively. For a square tube (Fig. A2.1)

$$b_{eq} = b \left[1 - \left(2.45 \frac{b}{t} - 50 \right) \left(\frac{r}{b} \right)^3 \right] \quad \text{(A2.9)}$$

and the moment of inertia

$$I_x = \frac{(b - 2r)^3 \, t}{6} + \frac{(b - 2r) \, t b^2}{2} + r^3 \, t\pi + \frac{(b - 2r)^2 \, r t\pi}{2} + 4(b - 2r) r^2 \, t$$

$$\text{(A2.10)}$$

where $b = B - t$. For cold-finished tubes, according to ISO/DIS 4019.2, $r/t = 2$, for hot-finished ones, according to ISO 657/XIV, $r/t = 1.125$.

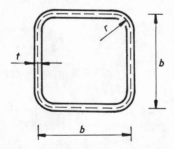

Fig. A2.1 – Dimensions of a cold- or hot-finished square hollow section

The effect of interaction may be considered by calculating the overall buckling strength with a reduced slenderness ratio

$$\overline{\lambda}' = \frac{\lambda}{\pi} \sqrt{\frac{R_y}{E} \overline{N}_v} \quad \text{(A2.11)}$$

$$\overline{N}' = \frac{1}{2\overline{\lambda}'^2} \left[1 + \eta_b + \overline{\lambda}'^2 - \sqrt{(1 + \eta_b + \overline{\lambda}'^2)^2 - 4\overline{\lambda}'^2} \right]. \quad \text{(A2.12)}$$

The overall buckling constraint is given by

$$\frac{N}{A} \leqslant R_y \, \overline{N}' \, \overline{N}_v \, .$$

Usami and Fukumoto (1982) have proposed another local buckling strength factor:

$$Q_F = \frac{0.75}{\overline{\lambda}_\nu} \leqslant 1.0 \qquad (A2.14)$$

and the overall buckling constraint in the form

$$\frac{N}{A} \leqslant R_y \, \overline{N}' \, Q_F \qquad (A2.15)$$

where $\overline{N}'$ is to be calculated by (A2.12) in which $\overline{\lambda}' = \overline{\lambda}\sqrt{Q_F}$.

A3 Buckling of Plates

A3.1 Effective Width of Compressed Plates

The critical elastic buckling stress of a compressed plate without initial imperfections and residual stresses is

$$\sigma_{cr} = \frac{k\pi^2 E}{12(1-\nu^2)} \left(\frac{t}{b}\right)^2 \qquad (A3.1)$$

where b and t are the width and thickness of the plate, respectively. In the case of simply supported edges $k = 4.0$, for steels $\nu = 0.3$. With the notation $\vartheta = b/t$ Eq. (A3.1) can be written as

$$\sigma_{cr} = 3.6152\, E/\vartheta^2 \; . \qquad (A3.2)$$

Above σ_{cr} the stress distribution is not uniform (Fig. A3.1). The post-critical behaviour of the plate may be described by means of the effective width b_e:

$$b_e \, \sigma_{max} = b \, \sigma_{av} \qquad (A3.3)$$

Assuming that (A3.2) is also valid for $\sigma_{max} - b_e$, and introducing the notation $\psi = b_e/b$ we get

$$\sigma_{max} = 3.6152\, E(t/b_e)^2 = 3.6152\, E/(\vartheta\,\psi)^2 \; ,$$

from which

$$\psi = 1.9014/\lambda_p; \qquad \lambda_p = \vartheta\sqrt{\sigma_{max}/E} \; , \qquad (A3.4)$$

λ_p denoting the plate slenderness. (A3.4) is valid provided the behaviour is elastic, i. e. if

$$3.6152\, E/\vartheta^2 \leqslant R_e \; . \qquad (A3.5)$$

$R_e = r R_y$ is the structural proportional limit in compression; for the base material $r = 0.75-0.80$, while for welded structural parts $r = 0.5-0.6$. Dividing

(A3.5) by $\sigma_{\max}$ and rearranging the inequality one obtains

$$\lambda_p \geqslant \lambda_{p0} = 1.9014\sqrt{\sigma_{\max}/R_e} \ . \tag{A3.5a}$$

For plates with initial imperfections and residual welding stresses an empirical formula proposed by Faulkner *et al.* (1973) can be used instead of (A3.4):

$$\psi = \frac{2}{\lambda_p} - \frac{1}{\lambda_p^{\,2}} - \frac{\sigma_c(\vartheta)}{R_y} \tag{A3.6}$$

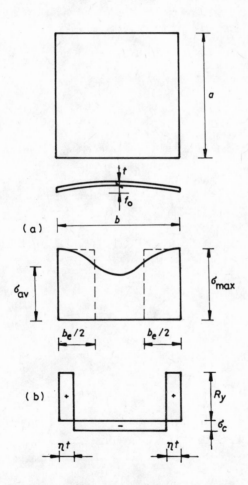

Fig. A3.1 − A simply supported rectangular plate loaded in compression
on two opposite edges of length b.
(a) Stress distribution in the post-buckling region;
(b) Simplified residual welding stress distribution due to longitudinal edge welds

where σ_c is the residual compressive stress. According to Appendix A1, in the case of the residual stress distribution shown in Fig. A1.2, we have

$$\frac{\sigma_c}{R_y} = \frac{2\eta}{\vartheta - 2\eta} . \qquad (A3.7)$$

η can be calculated from (A1.17) or can simply be taken as $\eta = 3$ and $\eta = 4.5$ for lightly and heavily welded parts, respectively. Instead of the R_y for the base material, the weld metal yield stress R_{y1} may be used, if the latter value is available (see Appendix A1).

Faulkner's formulae for the plastic zone, based on the tangent modulus, result in fairly intricate expressions. Since the formulae are empirical, the author has proposed a second-degree parabola (Farkas 1977c):

$$\psi = 1 - (1 - \psi_0)\left(\frac{\lambda_p}{\lambda_{p0}}\right)^2 ; \qquad \psi_0 = \frac{2}{\lambda_{p0}} - \frac{1}{\lambda_{p0}^2} - \frac{\sigma_c(\vartheta_0)}{R_y} \qquad (A3.8)$$

where

$$\vartheta_0 = 1.9014\, E/R_e .$$

In Fig. A3.2 some characteristic plate buckling curves are illustrated: a — without initial imperfections and residual stresses (von Kármán formula, Eq. (A3.4)); b — and c — with initial imperfections, with modified Faulkner's formulae (Eqs (A3.6)–(A3.8)); b — without residual stresses; c — with residual stresses, $\eta = 3$, $R_y = 245$ MPa, $E = 201$ GPa.

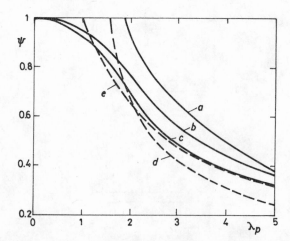

Fig. A3.2 – Plate buckling curves: effective width ratio versus plate slenderness.
$a - f_0 = 0$, $\sigma_c = 0$ (Eq. (A3.4)); $b - f_0 \neq 0$, $\sigma_c = 0$ (Eqs (A3.6) and (A3.8));
$c - f_0 \neq 0$, $\sigma_c \neq 0$, $\eta = 3$, $r = 0.6$, $\sigma_{max} = R_y$ (Eqs (A3.6) – (A3.8));
d – according to Braham *et al.* (1980), Eq. (A2.8), $\alpha_y = 0.67$; $e - (K_c^u + K_c^r)/2$
curve according to the BS 5400:Part 3

For the purpose of comparison two more curves are illustrated with dotted lines: d – the curve according to Braham *et al.* (1980)(A2.8), $\alpha_v = 0.67$, $\overline{\lambda}_{v0} = 0.8$; e – the $(K_c^u + K_c^r)/2$ curve according to the BS 5400: Part 3 (1982) for unrestrained (K_c^u) and restrained (K_c^r) plates, respectively.

It is worth noting that Vilnay and Rockey (1981) have derived generalised effective width formulae considering the effect of initial out-of-flatness and residual welding stresses in explicit form. Unfortunately, these formulae are valid only for $\lambda_p < 2.5$. Comparing graphically the modified Faulkner's curve with those of Vilnay – Rockey, one can conclude that in the modified Faulkner's formulae an initial out-of-flatness ratio of $f_0/t \geqslant 0.145\,\lambda_p$ is considered, which is suitable for design purposes. The same conclusion can be drawn by using curves published by Frieze *et al.* (1977).

A3.2 Limiting Values of Plate Slenderness

In order to make the designer's computational work easier, limiting slenderness values can be defined for plate elements of welded I- and box-beams (and also in the case of stiffened plates), below which the effective width ratio is $\psi \cong 1$. If the effect of initial imperfections and residual stresses may be neglected, this limiting slenderness can be calculated by (A3.1):

$$\sigma_{cr} = \frac{k\pi^2 E}{12(1 - \nu^2)\,\vartheta^2} = R_u .$$ (A3.9)

For *webs* of welded I- and box-beams, under bending action, we have $k = 23.9$, $\nu = 0.3$, $E = 210$ GPa, $R_u = 200$ MPa   (for steel 37), so that (A3.9) gives $\vartheta_{max} = 150$.

On the basis of experiments with box-beam models, Frieze (1980) found that the limiting value is $\vartheta_{max} = 145$. For *compressed flanges* of box-beams Frieze has given the value $\vartheta_{max} = 30$. It should be noted that this value is a little conservative for elastic design purposes. Some standards recommend a value of around 40. However, we propose the more conservative value of 30, because, according to *Frieze*, if $\vartheta_{flange} > 30$ and $\vartheta_{web} < 145$, or if $\vartheta_{fl} < 30$ and $\vartheta_w > 145$, the box-section cannot be treated as being fully active. Taking $k = 4.0$, Eq. (A3.9) gives $\vartheta_{max} = 61$. It can be seen that, because of the effect of residual welding stresses, this classical value should be reduced to 30.

For stress levels $\sigma_{max} < R_u$ and for other types of steel having R_u' instead of R_u these values should be reduced in the following manner:

$$\vartheta_{max} = 30\sqrt{\frac{R_{u37}}{\sigma_{max}}}\sqrt{\frac{R_{u37}}{R_u'}} .$$ (A3.10)

For Al-alloys the relevant value should also be multiplied by $\sqrt{E_{al}/E_s}$.

For *plastic design* these limiting values should be reduced in order to ensure the formation of plastic hinges without local buckling. The limiting values are given in Table A3.1.

Table A3.1

Limiting Values of Elastic and Plastic Plate Slendernesses for Steel 37

Structural part	Subject to	Elastic	Plastic
Web $1/\beta$, $1/\beta_p$	bending	145	70
	shear	90	70
	compression	40	43
Flange of I-beams $1/\delta$, $1/\delta_p$	compression	30	17
Flange of box beams $1/\delta$, $1/\delta_p$	compression	30–40	32

For *webs* subjected to bending, compression and shear, the following approximate interaction elastic design formula can be proposed (valid only for compressive σ_N):

$$\left(\frac{h}{t_w} \right)_{\text{lim}} = \frac{1}{\beta} \sqrt{k_{\text{red}}} \sqrt{\frac{\pi^2 E}{12(1 - \nu^2)R_u}} \tag{A3.11}$$

$$k_{\text{red}} = \sqrt{\frac{(\sigma_M + \sigma_N)^2 + 3\tau^2}{(\sigma_M/k_M)^2 + (\sigma_N/k_N)^2 + (\tau/k_\tau)^2}} \ . \tag{A3.12}$$

With values of $E = 210$ GPa, $R_u = R_{u37} = 200$ MPa, $\nu = 0.3$, Eq. (A3.11) takes the form

$$1/\beta = 30.81997 \sqrt{k_{\text{red}}} \ . \tag{A3.13}$$

Table A3.2

Limiting Plastic Plate Slendernesses $(1/\beta_p)$ for Various Steels for the Case of Combined Bending and Compression; $\nu_u = N/(AR_u)$

Steel	37	45	52
$0 \leqslant \nu_u \leqslant 0.27$	$70 - 100\,\nu_u$	$62 - 89\,\nu_u$	$57 - 81\,\nu_u$
$0.27 \leqslant \nu_u \leqslant 1$	43	38	35

For values $1/\beta = 145, 40$ and 90 (denoting bending, compression and shear, respectively) we get, from (A3.13), $k_M = 22.13$, $k_N = 1.684$ and $k_\tau = 4.903$.

With these values one obtains

$$\frac{1}{\beta} = 145 \sqrt[4]{\frac{(1 + \sigma_N/\sigma_M)^2 + 3(\tau/\sigma_M)^2}{1 + 173(\sigma_N/\sigma_M)^2 + 20(\tau/\sigma_M)^2}} \, . \tag{A3.14}$$

In the case of *plastic design* of webs subjected to bending and compression the limiting values are given in Table A3.2 (see, for example, (Eur. Recommend. (1978)).

A3.3 Ultimate Shear Strength of Plate Girders

In order to simplify the calculations for web buckling, limiting shear stresses can be determined below which the effect of shear on the web buckling may be neglected. It should be noted that the following limiting shear stress formulae are only valid if neither the buckling of compression flanges nor the lateral buckling of the beam occurs.

For hybrid I-beams without stiffeners (unstiffened, simply supported web) subjected to bending and shear Carskaddan (1968) has proposed the following formula

$$\tau_{cr} = \frac{5.34 \, \pi^2 E}{12(1 - \nu^2)} \left(\frac{t_w}{h}\right)^2 [1 - 1.61(1 - \xi_h)^2] \tag{A3.15}$$

where $\xi_h = R_{yw}/R_{uf}$, R_{yw} is the yield stress of the web, and R_{uf} is the ultimate stress of the flanges. Figure A3.3 shows the limiting shear stress curves for beams with a web made of steel 37, and adopting a safety factor $\gamma_b = 1.08$.

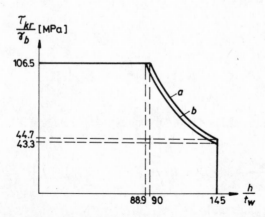

Fig. A3.3 – Limiting shear stress curves for unstiffened beams.
a – homogeneous I-beams made of steel 37; *b* – 52/37 hybrid I-beam,
$\gamma_b = 1.08$; $\tau_{uw} = 115$ MPa, $\xi_h = 0.86$

For homogeneous girders with unstiffened web made of steel 37 Delesques (1979) has proposed an empirical formula

$$\tau_{cr} \text{ [MPa]} = 16 + 8889 \, t_w/h \qquad (A3.16)$$

for $\quad 67 < h/t_w < 167$.

Bending stresses should be considered by an interaction formula only if $h/t_w > 167$.

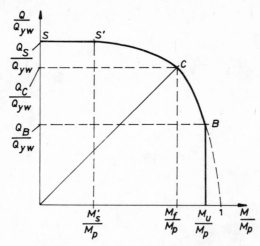

Fig. A3.4 – Interaction diagram for plate girders with transverse stiffeners

For homogeneous and hybrid I-beams with transverse stiffeners the method of Rockey *et al.*(1978) can be applied. The characteristic points of the interaction diagram shown in Fig. A3.4 may be calculated by means of the following formulae:

$$M_p = R_{yf} \, bt_f h + R_{yw} h^2 \, t_w/4 = M_f + M_w \qquad (A3.17)$$

$$M_u = R_{uf} \left[bt_f h + \frac{h^2 \, t_w}{12} (3\xi_h - \xi_h^3) \right]. \qquad (A3.18)$$

For the point S' we obtain $Q_S' = Q_S$ and $M_S' = Qb$ (i. e. the maximum bending moment in the end panel of a simply supported beam) with the restriction that $M_S' \leqslant 0.5 \, M_f$.

$$\frac{Q_S}{Q_{yw}} = \frac{\tau_{cr}}{\tau_{yw}} + 3 \sin^2 \theta \left(\cot \theta - \frac{a}{h} \right) \frac{\sigma_t^y}{R_{yw}} + 4\sqrt{3} \sin \theta \sqrt{\frac{\sigma_t^y}{R_{yw}} \, M_p^*} \quad (A3.19)$$

where

$$\frac{\sigma_t^y}{R_{yw}} = -\frac{\sqrt{3} \, \tau_{cr}}{2 \, \tau_{yw}} \sin \theta + \sqrt{1 + \left(\frac{\tau_{cr}}{\tau_{yw}} \right)^2 (0.75 \sin^2 2\theta - 1)} \quad (A3.20)$$

and

$$M_p^* = M_{pf}/(R_{yw} h^2 t_w) \, ; \qquad M_{pf} = R_{yf} bt_f^2/4 \qquad \text{(A3.21)}$$

$$Q_{yw} = \tau_{yw} ht_w = \frac{R_{yw}}{\sqrt{3}} ht_w \qquad \text{(A3.22)}$$

$$\tau_{cr} = \frac{k_\tau \pi^2 E}{12(1-\nu^2)} \left(\frac{t_w}{h}\right)^2 \qquad \text{(A3.23)}$$

$$k_\tau = 5.34 + 4(h/a)^2 \qquad \text{for} \qquad a/h \geqslant 1$$

$$k_\tau = 4.0 + 5.34(h/a)^2 \qquad \text{for} \qquad a/h < 1 \, .$$

The second and third terms in (A3.19) express the effect of the membrane tension field and the flexural rigidity of the flanges, respectively.

$$\frac{Q_C}{Q_{yw}} = \frac{\tau_{cr}}{\tau_{yw}} + \frac{\sigma_t^y}{R_{yw}} \sin\left(\frac{4\theta_d}{3}\right) \left[2 - \left(\frac{a}{h}\right)^{1/8}\right] \left(0.544 + \frac{36.8 M_{pf}}{M_f}\right) . \text{(A3.24)}$$

The section B–C of the interaction diagram may be represented by a simple parabola, so that

$$\frac{Q_B}{Q_C} = \sqrt{\frac{M_p - M_u}{M_w}} \qquad \text{(A3.25)}$$

$$\theta = 2\theta_d/3 \, ; \qquad \tan \theta_d = h/a \qquad \text{(A3.26)}$$

where a is the distance between the transverse stiffeners.

According to the Swiss standard SIA 161 (1979) the limiting shear force of an *unstiffened homogeneous I-beam* subjected to a bending moment M is

$$Q_{cr} = \tau_{cr} ht_w \qquad \text{if} \qquad M < M_f \, .$$

τ_{cr} is to be calculated from (A3.15) ($\xi_h = 1$). If $M > M_f$, the limiting shear force is given by

$$Q = Q_{cr} \sqrt{\frac{M_p - M}{M_w}} \, . \qquad \text{(A3.27)}$$

This agrees with formulae (A3.24) and (A3.25) given above. In order to obtain the ultimate bending moment in the case of an elastic stress distribution and $\sigma_{max} = R_y$, the compressed part of the web may be considered with only an effective width active, provided $h/t_w > 125$ (for 37-steels).

A4 Static and Dynamic Response of Sandwich Beams with Outer Layers of Box Cross-Section

In order to improve the low vibration damping capacity of metal structures, it is advantageous to apply layers of high damping materials, e. g. rubber, polymers, foams. Yin *et al.* (1967) have shown that maximum damping can be achieved by sandwich beams made of two rectangular tubes and a constrained damping layer glued between them. Farkas and Jármai (1982) have worked out a structural synthesis for beams of this type.

The static behaviour of such sandwich beams can be treated on the basis of Allen's (1973) paper in which the following assumptions are made: (1) the stiffness of the core may be neglected, i. e. normal stresses do not occur in the core while the shear stresses in the core cross-section are constant; (2) the transverse strain of the core can be neglected; (3) the shear deformations of the faces may be ignored.

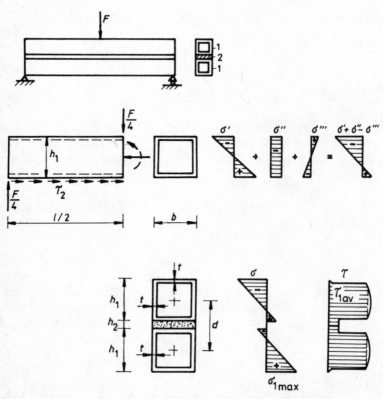

Fig. A4.1 – Stress diagrams for a symmetrical sandwich beam with flexurally stiff outer layers

In the case of the sandwich beam shown in Fig. A4.1 the maximum deflection is

$$w_{max} = \frac{Fl^3}{48\,B} + \frac{Fl}{4\,B_q}\left(1 - \frac{B_f}{B}\right)\left(1 - \frac{\tanh \chi}{\chi}\right) \qquad \text{(A4.1)}$$

where

$$\chi = \frac{1}{2}\left[\frac{B_f}{B_q\,l^2}\left(1 - \frac{B_f}{B}\right)\right]^{-1/2} \qquad \text{(A4.2)}$$

$$B = B_f + B_s\;; \qquad B_f = 2E_1\,I_1\;; \qquad B_s = E_1\,A_1\,d^2/2\;; \qquad \text{(A4.3)}$$

$$B_q = G_s\,bd^2/h_2\;. \qquad \text{(A4.4)}$$

G_s being the static shear modulus of the core material;

$$I_1 = \frac{t_1(h_1 - 2t_1)^3}{6} + bt_1\,h_1^2/2\;; \qquad \text{(A4.5)}$$

$$A_1 = 2t_1(h_1 - 2t_1) + 2bt_1\;. \qquad \text{(A4.6)}$$

The maximum shear stress in the core is given by

$$\tau_2 = \frac{F}{2b(h_1 + h_2)}\left(1 - \frac{1}{\cosh \chi}\right)\;. \qquad \text{(A4.7)}$$

The expression for the maximum normal stress is

$$\sigma_{1\,max} = \frac{E_1\,Fl}{4}\left[\frac{1}{B}\left(h_1 + \frac{h_2}{2}\right)\left(1 - \frac{\tanh \chi}{\chi}\right) + \frac{h_1\tanh \chi}{2B_f\chi}\right]\;. \qquad \text{(A4.8)}$$

Formulae were also derived for a uniformly distributed load by Allen (1973), and for the case of two concentrated forces by Grosskopf and Winkler (1973).

The diagrams of normal and shear stresses are shown in Fig. A4.1. From the equilibrium of a quarter of the beam it can be seen that the following stress components occur:

$$\sigma' = \frac{Fl}{8\,W_1}\;; \qquad \sigma'' = \frac{\tau_2\,bl}{2\,A_1}\;; \qquad \sigma''' = \frac{\tau_2\,blh_1}{4\,W_1}\;. \qquad \text{(A4.9)}$$

The mean shear stress in the webs of the face tubes can be estimated approximately by the expression

$$\tau_{1m} = \frac{F/2 - \tau_2\,bh_2}{4\,t(h_1 - 2\,t_1)}\;. \qquad \text{(A4.10)}$$

In the case of *dynamic analysis*, the following differential equation was derived by Mead and Markuš (1969) for the forced vibratory motion of a sandwich beam:

$$\frac{\partial^6 w}{\partial x^6} - g_0 (1 + Y) \frac{\partial^4 w}{\partial x^4} = \frac{1}{B_f} \left(\frac{\partial^4 p_0}{\partial x^4} - g_0 p_0 \right) \qquad (A4.11)$$

where

$$p_0 = -m \frac{\partial^2 w}{\partial \overline{t}^2} + p(x, \overline{t}) .$$

For sandwich beams with symmetric cross-section we also have

$$g_0 = \frac{2 G_d b}{h_2 A_1 E} \qquad (A4.12)$$

and

$$Y = \frac{(h_1 + h_2)^2}{2 B_f} A_1 E . \qquad (A4.13)$$

The eigenfrequencies and loss factors can be determined by means of diagrams given in the article of Markuš and Valášková (1972) or in the book of Markuš *et al.* (1977). Ungar (1962) has given the following formula for the loss factor

$$\eta_d = \frac{\eta_{d2} X Y}{1 + (2 + Y) X + (1 + Y)(1 + \eta_{d2}^2) X^2} \qquad (A4.14)$$

where

$$X = g_0 \left(\frac{C_d l^2}{2\pi} \right)^2 \qquad (A4.15)$$

is the shear parameter.

C_d is a constant relating to the vibration mode. For instance, according to Yin *et al.* (1967), for the first four eigenfrequencies of a two-pinned beam $C_d = 2; 1; 0.67; 0.50$.

It should be noted that the dynamic shear modulus G_d and the loss factor η_{d2} for the core material depend on the shear stress level.

Snowdon (1968) has shown that, for a simple one-mass system, and for $\eta_d \leqslant \leqslant 0.1$

$$\eta_d \approx 1/T_0 = |x_1/x_2| = |F_1/F_2| \qquad (A4.16)$$

where T_0 is the transmissibility at the resonance frequency; F_1 is the applied force; F_2 is the dynamic force, and x_1, x_2 are the corresponding displacements.

A5 Calculation of Cellular Plates by Means of the Theory of Orthotropic Plates

On the basis of the theory of plates (Timoshenko and Woinowsky-Krieger 1959), (Lee *et al.* 1971), the relationships between the in-plane strains and the derivatives of the transverse deflection w are as follows (Fig. A5.1):

$$\epsilon_x = -zw'' \; ; \qquad \epsilon_y = -zw^{\cdot\cdot} \; ; \qquad \gamma_{xy} = -2zw^{'\cdot} \qquad (A5.1)$$

where the prime $(')$ and dot $(\cdot)$ superscripts denote partial derivatives with respect to x and y respectively.

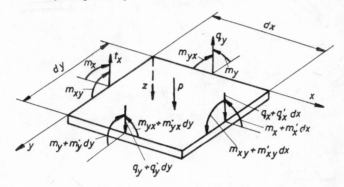

Fig. A5.1 – Equilibrium of a plate element

By introducing notation $E_1 = E/(1 - \nu^2)$, and, on the basis of (A5.1), we get the following formulae for the stress components:

$$\left. \begin{array}{l} \sigma_x = E_1(\epsilon_x + \nu\epsilon_y) = -E_1 z(w'' + \nu w^{\cdot\cdot}) \\[2mm] \sigma_y = -E_1 z(w^{\cdot\cdot} + \nu w''); \quad \tau_{xy} = -2Gzw^{'\cdot} \end{array} \right\} \qquad (A5.2)$$

where $G = E/[2(1 + \nu)]$ is the shear modulus.

The formulae for the bending and twisting moments per unit length are as follows:

$$\left. \begin{array}{l} m_x = \int \sigma_x z \, dA = -B_x(w'' + \nu w^{\cdot\cdot}) \\[2mm] m_y = -B_y(w^{\cdot\cdot} + \nu w'') \\[2mm] m_{xy} = \int \tau_{xy} z \, dA = 2B_{xy} w^{'\cdot} \; ; \quad m_{yx} = -2B_{yx} w^{'\cdot} \, . \end{array} \right\} \qquad (A5.3)$$

The bending and torsional stiffnesses of the plate are given by (Fig. A5.2(a)):

$$\left. \begin{array}{l} B_x = \dfrac{E}{a_y} \sum \dfrac{I_{yi}}{1 - \nu_i^2} \; ; \qquad B_y = \dfrac{E}{a_x} \sum \dfrac{I_{xi}}{1 - \nu_i^2} \\[4mm] B_{xy} = G I_y / a_y \; ; \qquad B_{yx} = G I_x / a_x \end{array} \right\} \qquad (A5.4)$$

where, for the face-sheets, $\nu_i = \nu$, and, for the ribs, $\nu_i = 0$. a_x, a_y are the spacings between the ribs, while I_x, I_y are the moments of inertia.

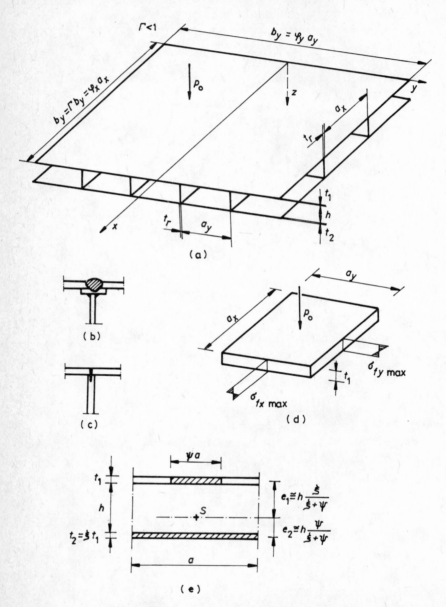

Fig. A5.2 – (a) Main dimensions of a rectangular cellular plate; (b) CO_2 arc-spot-welded joint between face-sheet and rib using a backing strip; (c) electron-beam-welded joint; (d) local bending of an upper face-sheet; (e) effective plate cross-section

From the equilibrium equations of a plate element (Fig. A5.1) one obtains

$$\left.\begin{array}{l} q_x = m'_x + \dot{m}_{yx} = -[B_x\, w''' + (2B_{yx} + \nu B_x)\,w'^{\cdots}] \\[2mm] q_y = \dot{m}_y - m'_{xy} = -[B_y\, w^{\cdots} + (2B_{xy} + \nu B_y)w'''^{\cdot}] \end{array}\right\} \qquad (A5.5)$$

and

$$q'_x + \dot{q}_y + p = 0. \qquad (A5.6)$$

Inserting (A5.5) into (A5.6) yields Huber's equation for orthotropic plates:

$$B_x\, w'''' + 2Hw''^{\cdots} + B_y\, w^{\cdots\cdots} = p(x, y) \qquad (A5.7)$$

where

$$H = B_{xy} + B_{yx} + \frac{\nu}{2}\,(B_x + B_y) \qquad (A5.8)$$

is the torsional stiffness of an orthotropic plate.

Neglecting the moment of inertia of the ribs and taking into account an effective width of the compressed face-sheet (Fig. A5.2(e)) we obtain

$$\begin{aligned} B_x &= \frac{E_1\, I_y}{a_y} = E_1\, t_1\, h^2\; \frac{\xi\,\psi_y}{\xi + \psi_y} \\[4mm] B_y &= \frac{E_1\, I_x}{a_x} = E_1\, t_1\, h^2\; \frac{\xi\,\psi_x}{\xi + \psi_x} \end{aligned} \qquad (A5.9)$$

where $\xi = t_2/t_1$. The effective width coefficients ψ_x and ψ_y can be calculated by means of the formulae presented in Appendix A3.

Stresses due to the local bending of the individual plate-elements of the upper face can be calculated by means of formulae valid for isotropic plates with clamped edges (see (Timoshenko and Woinowsky-Krieger 1959)) (Fig. A5.2(d)):

$$\begin{aligned} \sigma_{fy\,\max} &= \frac{c_{fy}\,\gamma_0\,p_0\,a_y^2}{t_1^2/6} \\[4mm] \sigma_{fx\,\max} &= 6\gamma_0\,p_0\,c_{fx}\,a_y^2/t_1^2 \end{aligned} \qquad (A5.10)$$

where γ_0 is the safety factor, while the coefficients c_{fy} and c_{fx} are given in Table A5.1.

By using (A5.4) and (A5.9) the torsional stiffness H can be written in the form

$$H = \frac{E_1}{2}\left(\frac{I_x}{a_x} + \frac{I_y}{a_y}\right). \qquad (A5.11)$$

Since, in the case of cellular plates,

$$\theta_H = \frac{H}{\sqrt{B_x\, B_y}} = \frac{1}{2}\,\frac{I_x/a_x + I_y/a_y}{\sqrt{(I_x/a_x)(I_y/a_y)}} \geqslant 1, \qquad (A5.12)$$

Table A5.1

Coefficients for the Calculation of Maximum Stresses in a Uniformly Loaded Rectangular Isotropic Plate with Clamped Edges

a_x/a_y	1.0	1.1	1.2	1.3	1.4	1.5	1.6	1.7	1.8	1.9	2.0	∞
$10^4\, c_{fy}$	513	581	639	687	726	757	780	799	812	822	829	833
$10^4\, c_{fx}$	513	538	554	563	568	570	571	571	571	571	571	571

Table A5.2

Coefficients for the Calculation of Maximum Deflection and Stresses in a Simply Supported Uniformly Loaded Rectangular Orthotropic Plate

θ_B	1.0	1.1	1.2	1.3	1.4	1.5	1.6	1.7	1.8	1.9	2.0	3.0	4.0	5.0	∞
$10^4\, c_{x\sigma}$	526	609	689	763	830	892	947	998	1042	1082	1117	1307	1357	1369	1374
$10^4\, c_{y\sigma}$	526	541	551	552	552	547	541	534	526	518	510	446	422	412	412
$10^5\, c_w$	406	485	564	638	705	772	830	883	931	974	1013	1223	1282	1297	1302

we can take $\theta_H = 1$. It is worth noting that this property of cellular plates was also confirmed by static torsional tests (Farkas 1974c).

The solution of (A5.7) for simply supported, uniformly loaded rectangular cellular plates for the special case $\theta_H = 1$ and $B_x \neq B_y$ can be obtained by using *Timoshenko's* solution for isotropic plates (see (Farkas 1976)). Instead of this exact solution the approximation $B_x = B_y$ suggested by Schade (1941) can also be used.

Neglecting the effect if shear deformations, the maximum deflection can be expressed as

$$w_{\max} = c_w \, \frac{p_d + p_0}{B_x} \, b_x^4 \, . \tag{A5.13}$$

The maximum normal stresses in the x and y directions for the upper face-sheet are given by

$$\sigma_{x\max 1} = c_{x\sigma} \, \frac{E p_1 \, b_x^2 \, e_{y1}}{B_x} \, ; \qquad \sigma_{y\max 1} = c_{y\sigma} \frac{E p_1 \, b_x^2 \, e_{x1}}{\sqrt{B_x B_y}} \, . \tag{A5.14}$$

The coefficients $c_{x\sigma}$, $c_{y\sigma}$ and c_w in the above expressions are given in Table A5.2 as a function of

$$\theta_B = \frac{b_y}{b_x} \sqrt[4]{\frac{B_x}{B_y}} = \frac{1}{\Gamma} \sqrt[4]{\frac{\psi_y(\xi + \psi_x)}{\psi_x(\xi + \psi_y)}} \, ; \qquad \Gamma = \frac{b_x}{b_y} \leqslant 1 \tag{A5.15}$$

while

$$p_1 = \gamma_d \, p_d + \gamma_0 \, p_0 \tag{A5.16}$$

where p_d is the intensity of the dead weight, and γ_d, γ_0 are safety factors.

In the case of an isotropic plate, the maximum edge load is (see (Timoshenko and Woinowsky-Krieger 1959))

$$Q_{\max} = c_q \, p_1 \, b_x \, . \tag{A5.17}$$

Since c_q varies within a small range ($c_q = 0.42 - 0.50$), we take $c_q = 0.50$ for rectangular plates and $c_q = 0.42$ for square plates. The mean shear stress in a rib is

$$\tau = \frac{c_q \, p_1 \, b_x \, a_x}{h t_r} \, . \tag{A5.18}$$

The eigenfrequencies of a simply supported rectangular cellular plate can be calculated on the basis of the equation (see, e. g., (Szilard 1974))

$$B_x \, w'''' + 2Hw'''\cdot + B_y \, w\cdots + m_{\text{red}} \, \frac{\partial^2 w}{\partial t^2} = 0 \tag{A5.19}$$

where

$$m_{\text{red}} = \rho \, t_{\text{red}}; \qquad t_{\text{red}} = t_1 + t_2 + t_r h \left(\frac{1}{a_x} + \frac{1}{a_y} \right).$$

Inserting into (A5.19) the function

$$w(x, y, \bar{t}) = w_{mn} \sin \frac{m \pi x}{b_x} \sin \frac{n \pi y}{b_y} \sin \omega \, \bar{t} \qquad (A5.20)$$

one obtains

$$\omega_{mn}^2 = \frac{\pi^4}{m_{\text{red}} \, b_x^4} [B_x \, m^4 + 2H(mn\Gamma)^2 + B_y(n\Gamma)^4]. \qquad (A5.21)$$

For a simply supported isotropic rectangular plate element of the upper face sheet we have

$$B_x = H = B_y = E_1 \, t_f^3 / 12 \, ; \qquad \Gamma = a_y/a_x$$

so that, for the first eigenfrequency $(m = n = 1)$, we get

$$\omega_{11f}^2 = \frac{E_1 \, t_f^2 \, \pi^4}{12 \, \rho \, a_y^4} \left[1 + \left(\frac{a_y}{a_x} \right)^2 \right]^2. \qquad (A5.22)$$

The cost function for a rectangular cellular plate consists of the material and welding costs. We assume that φ_x and φ_y are the numbers of ribs in x and y directions, respectively. The cost of the material can be expressed as follows:

$$K_1 = k_1 \, \rho[\cdot b_x \, b_y(t_1 + t_2) + t_r \, h(\varphi_x \, b_y + \varphi_y \, b_x)] \qquad (A5.23)$$

where k_1 [\$/kg] is the material cost factor.

The cost of the welding consists of two parts:

$$K_2 = 4k_2 \, h \, \varphi_x \, \varphi_y + (2k_3 + k_4 + k_5)(\varphi_x \, b_y + \varphi_y \, b_x). \qquad (A5.24)$$

The first term expresses the welding cost of the four fillet welds connecting the ribs at the intersections while the second term describes the cost of welds between the face-sheets and ribs. k_3 is the cost factor of welds which connect the backing strips (Fig. A5.2(b)) to the ribs; k_4 and k_5 are the cost coefficients of the welds between the backing strips and the face-sheet at the lower and upper side of the plate, respectively. It can be assumed that $k_5/k_4 = t_1/t_2 = 1/\xi$.

A6 Fortran Program of the Backtrack Programming Method

```
PROGRAM TITLE = OPTIMUM ELASTIC DESIGN OF A SINGLE-BAY
      PITCHED-ROOF PORTAL FRAME WITH EIGHT VARIABLES BY
      BACKTRACK PROGRAMMING
PROGRAM  CODE NAME = MISZABOO
WRITER = DR LASZLO SZABO
ORGANIZATION AND AVAILABILITY = TECHN. UNIV. MISKOLC DEPT
      OF MATERIALS HANDLING EQUIPMENTS PROF DR JOZSEF FARKAS

        COMMON H1,VG1,S1,V1,H2,VG2,S2,V2,A,B,NN,VMIN,SW
        DIMENSION VV(9)
        K = 9
        KV = 7
        A = 760.
        B = 1305.
        HD = 2.0
        VGD = 0.1
        SD = 2.0
        IDVV = 1
C
C       H     =  WEB HEIGHT
C       VG    =  WEB THICKNESS
C       S     =  FLANGE WIDTH
C       V     =  FLANGE THICKNESS
C       SECT  =  NUMBER OF ALL COMBINATIONS
C       VKER  =  HALF VOLUME OF THE FRAME
C       NN    =  NUMBER OF TESTED COMBINATIONS
C
        READ(5,200) (VV(I),I = 1,K)
        NN = 0
        SW = 0.
        READ(5,20) H1A,H1F,S1A,S1F,H2A,H2F,S2A,S2F
        CALL PRT (H1A,3HH1A)
        CALL PRT (H1F,3HH1F)
        CALL PRT (S1A,3HS1A)
        CALL PRT (S1F,3HS1F)
        CALL PRT (H2A,3HH2A)
        CALL PRT (H2F,3HH2F)
        CALL PRT (S2A,3HS2A)
        CALL PRT (S2F,3HS2F)
        CALL PRT (VV(1),5HVV(1))
        CALL PRT (VV(2),5HVV(2))
        CALL PRT (VV(3),5HVV(3))
```

```
          CALL PRT (VV(4),5HVV(4))
          CALL PRT (VV(5),5HVV(5))
          CALL PRT (VV(6),5HVV(6))
          CALL PRT (VV(7),5HVV(7))
          CALL PRT (VV(8),5HVV(8))
          CALL PRT (VV(9),5HVV(9))
          CALL PRT (HD,2HHD)
          CALL PRT (SD,2HSD)
          SECT = ((H1F−H1A)/HD+1.)*((S1F−S1A)/SD+1.)*((H2F−H2A)/HD+
          11.)*((S2F−S2A)/SD+1.)*(KV**4)
          CALL PRT (SECT,4HSECT)
          WRITE (6,21)
          HD = 8.
          SD = 8.
          START = 0.
          GOTO 1010
1000      CONTINUE
          CALL LIMIT (H1A,H1F,H1M,HD)
          CALL LIMIT (S1A,S1F,S1M,SD)
          CALL LIMIT (H2A,H2F,H2M,HD)
          CALL LIMIT (S2A,S2F,S2M,SD)
          HD = 2.
          SD = 2.
          START = 1.
          CALL PRT (START,5HSTART)
1010      CONTINUE
          H1 = H1F
          VG1 = VV(K)
          S1 = S1F
          V1 = VV(K)
          H2 = H2F
          VG2 = VV(K)
          S2 = S2F
          V2 = VV(K)
          IF (START.EQ.1.) GOTO 1020
          VMIN = 1.E8
          CALL GRD (GRDN, VKER)
          IF (GRDN.LT.0.5) GOTO 22
          VMIN = VKER
1020      CONTINUE
          CALL HALVING (H1,H1A,H1F,HD)
1         CONTINUE
          S1 = S1F
          V1 = VV(K)
```

```
        H2 = H2F
        VG2 = VV(K)
        S2 = S2F
        V2 = VV(K)
        CALL IHALVING(VG1,IDVV,IVG1,K,VV)
2       CONTINUE
        V1 = VV(K)
        H2 = H2F
        VG2 = VV(K)
        S2 = S2F
        V2 = VV(K)
        CALL HALVING (S1,S1A,S1F,SD)
3       CONTINUE
        H2 = H2F
        VG2 = VV(K)
        S2 = S2F
        V2 = VV(K)
        CALL IHALVING (V1,IDVV,IV1,K,VV)
4       CONTINUE
        VG2 = VV(K)
        S2 = S2F
        V2 = VV(K)
        CALL HALVING (H2,H2A,H2F,HD)
5       CONTINUE
        S2 = S2F
        V2 = VV(K)
        CALL IHALVING (VG2,IDVV,IVG2,K,VV)
6       CONTINUE
        V2 = VV(K)
        CALL HALVING (S2,S2A,S2F,SD)
7       CONTINUE
        CALL MINV2 (IDVV,IV2,K,VV)
        IF (V2.LT.VV(1)) GOTO 13
        IF (V2.EQ.VV(K)) GOTO 11
        GOTO 10
8       CONTINUE
        SW = 1.
        CALL GRD (GRDN,VKER)
        SW = 0.
        IF (VKER.GT.VMIN) GOTO 9
        H1M= H1
        VG1M= VG1
        S1M = S1
        V1M = V1
```

```
        H2M = H2
        VG2M = VG2
        S2M = S2
        V2M = V2
        VMIN = VKER
  9     IF (S2.GT.S2F−0.01) GOTO 13
        S2 = S2+SD
        CALL MINV2 (IDVV,IV2,K,VV)
        IF (V2.LT.VV(1)) GOTO 13
        IF (V2.EQ.VV(K)) GOTO 11
 10     CONTINUE
        CALL GRD (GRDN,VKER)
        IF (GRDN.LT.0.5) GOTO 9
 11     CONTINUE
        IF (IV2.EQ.1) GOTO 8
110     IV2 = IV2−IDVV
        V2=VV(IV2)
        IV2D=IV2+IDVV
        IF(VV(IV2).EQ.VV(IV2D)) GOTO 110
        CALL GRD (GRDN,VKER)
        IF (GRDN.LT.0.5) GOTO 12
        GOTO 11
 12     IV2=IV2+IDVV
        V2=VV(IV2)
        GOTO 8
 13     CONTINUE
        IF (VG2.EQ.VV(K)) GOTO 14
        IVG2=IVG2+IDVV
        VG2=VV(IVG2)
        GOTO 6
 14     IF (H2.GT.H2F−0.01) GOTO 15
        H2=H2+HD
        GOTO 5
 15     IF (V1.EQ.VV(K)) GOTO 16
        IV1=IV1+IDVV
        V1=VV(IV1)
        GOTO 4
 16     IF (S1.GT.S1F−0.01) GOTO 17
        S1=S1+SD
        GOTO 3
 17     IF (VG1.EQ.VV(K)) GOTO 18
        IVG1=IVG1+IDVV
        VG1=VV(IVG1)
        GOTO 2
```

```
18     IF (H1.GT.H1F−0.01) GOTO 19
       H1=H1+HD
       GOTO 1
19     CONTINUE
       IF (START.LT.0.5) GOTO 1000
       CALL PRT (H1M,2HH1)
       CALL PRT (VG1M,3HVG1)
       CALL PRT (S1M,2HS1)
       CALL PRT (V1M,2HV1)
       CALL PRT (H2M,2HH2)
       CALL PRT (VG2M,3HVG2)
       CALL PRT (S2M,2HS2)
       CALL PRT (V2M,2HV2)
       CALL PRT (VMIN,4HVMIN)
       AN=FLOAT(NN)
       CALL PRT (AN,2HNN)
20     FORMAT (8F9.2)
200    FORMAT (9F8.2)
21     FORMAT (/////3X,2HH1,5X,3HVG1,4X,2HS1,6X,2HV1,4X,2HH2,5X,
       13HVG2,4X,2HS,6X,2HV2,4X,4HGRDN,7X,4HVKER,10X,4HVMIN,
       27X,3HFR1,5X,3HFR2,6X,2HNN,//)
22     CONTINUE
       STOP
       END

       SUBROUTINE GRD (GRDN,VKER)
C
C      CALCULATION OF THE VALUE OF OBJECTIVE FUNCTION AND
C      CHECKING THE FULFILMENT OF CONSTRAINTS
C
       REAL I1,I2,NA,MA,K1,K2,MC,NC2,MC2
       COMMON H1,VG1,S1,V1,H2,VG2,S2,V2,A,B,NN,VMIN,SW
       GRDN=0.
       FR2=0.
       NN=NN+1
       FH=20.
       NA=116.6
       VKER=1.E8
       A1=H1*VG1+2.*S1*V1
       A2=H2*VG2+2.*S2*V2
       K1=VG1*H1**2/6.+S1*V1*H1
       K2=VG2*H2**2/6.+S2*V2*H2
       I1=K1*H1/2.
       I2=K2*H2/2.
       OMA=I2/I1
```

13*

```
1    CONTINUE
     OMA1=326689.*OMA
     OMA2=1.0175*OMA**2
     OMA3=13.86245*OMA
     OMAX=OMA2+OMA3+0.73031
     MA=(OMA1+98959.)/OMAX
     FM1=MA/K1
     FN1=NA/A1
     FR1=FM1+FN1
     IF (FR1.GT.FH) GOTO 6
2    CONTINUE
     OMA4=496384.*OMA
     MC=(OMA4+8866.)/OMAX
     HA=(MA+MC)/A
     NC2=28.373+0.957826*HA
     MC2=817.47*HA−MA−20626.
     FM2=MC2/K2
     FN2=NC2/A2
     FR2=FM2+FN2
     IF (FR2.GT.FH) GOTO 6
3    CONTINUE
     X1=(FN1/FM1)
     X11=(1.+X1)**2
     X12=1.+173.*X1**2
     C1M=145.*SQRT(SQRT(X11/X12))
     C1=H1/VG1
     IF(C1.GT.C1M) GOTO 6
4    CONTINUE
     X2=(FN2/FM2)
     X21=(1.+X2)**2
     X22=1.+173.*X2**2
     C2M=145.*SQRT(SQRT(X21/X22))
     C2=H2/VG2
     IF (C2.GT.C2M) GOTO 6
5    CONTINUE
     C3=S1/V1
     C3M=30.0001
     IF (C3.GT.C3M) GOTO 6
     C4=S2/V2
     IF (C4.GT.C3M) GOTO 6
     GRDN=1.
     A1=H1*VG1+2.*S1*V1
     A2=H2*VG2+2.*S2*V2
     VKER=A*A1+B*A2
```

```
      6     CONTINUE
            IF (SW.LT.0.5) GOTO 8
            WRITE (6, 7) H1,VG1,S1,V1,H2,VG2,S2,V2,GRDN,VKER,VMIN,FR1,
           1FR2,NN
      7     FORMAT (9(2X,F5.2),2(2X,F12.2),2(2X,F6.2),I8)
      8     RETURN


            SUBROUTINE HALVING (X,XA,XF,XD)
C
C           INTERVAL HALVING METHOD
C
            COMMON H1,VG1,S1,V1,H2,VG2,S2,V2,A,B,NN,VMIN,SW
            X=XA
            CALL GRD (GRDN,VKER)
            IF (GRDN.GT.0.5) GOTO 5
            XI=(XF−XA)/2.
            X=XF
      1     CONTINUE
            X=X−XI
      2     CALL GRD (GRDN,VKER)
            IF (XI.LT.XD+0.01) GOTO 4
            XI=XI/2.
            IF (GRDN.LT.0.5) GOTO 3
            GOTO 1
      3     X=X+XI
            GOTO 2
      4     IF (GRDN.LT.0.5) X=X+XI
      5     RETURN
            END


            SUBROUTINE IHALVING (VVX,IDVV,J,K,VV)
C
C           INTERVAL HALVING METHOD WITH BASIC SERIES OF INTEGER
C           VALUES
C
            DIMENSION VV(K)
            COMMON H1,VG1,S1,V1,H2,VG2,S2,V2,A,B,NN,VMIN,SW
            VVX=VV(1)
            J=1
            CALL GRD (GRDN,VKER)
            IF (GRDN.GT.0.5) GOTO 5
            IX=(K−1)/2
            I=K
      1     CONTINUE
            I=I−IX
```

```
          VVX=VV(I)
          IIX=I+IX
          IF (VV(I).EQ.VV(IIX)) GOTO 4
    2     CONTINUE
          CALL GRD (GRDN,VKER)
          IF (IX.EQ.IDVV) GOTO 4
          IX=IX/2
          IF (GRDN.LT.0.5) GOTO 3
          GOTO  1
    3     I=I+IX
          VVX=VV(I)
          GOTO  2
    4     IF (GRDN.LT.0.5) I=I+IX
          VVX=VV(I)
          J=I
    5     RETURN
          END


          SUBROUTINE MINV2 (IDVV,I,K,VV)
C
C         CALCULATION OF THE 8TH VARIABLE FROM THE OBJECTIVE
C         FUNCTION  WITH GIVEN VALUES OF THE OTHER 7
C         VARIABLES
C
          DIMENSION VV(K)
          COMMON H1,VG1,S1,V1,H2,VG2,S2,V2,A,B,NN,VMIN,SW
          A1=H1*VG1+2.*S1*V1
          AA1=A*A1
          HVG=H2*VG2
          S22=2.*S2
          V2=((VMIN−AA1)/B−HVG)/S22
          I=K
          IF (V2.GT.VV(K)) GOTO 3
          IF (V2.LT.VV(1)) GOTO 4
          DO 1 I=1,K,IDDV
          IF (VV(I)−V2) 1,2,20
    1     CONTINUE
    2     V2=VV(I)
          GOTO  4
   20     I=I−1
          V2=VV(I)
          GOTO  4
    3     V2=VV(K)
```

```
    4    RETURN
         END


         SUBROUTINE LIMIT (XA,XF,XM,XD)
C
C        DETERMINATION OF THE SIZE LIMITS NEAR THE OPTIMUM
C        FOUND IN FIRST STAGE
C
         IF (XM.EQ.XA) GOTO 1
         XA=XM−XD
         GOTO 2
    1    XA=XM
         XF=XA+XD
         GOTO 4
    2    IF (XM.EQ.XF) GOTO 3
         XF=XM+XD
         GOTO 4
    3    XF=XM
    4    RETURN
         END


         SUBROUTINE PRT (X,NAME)
C
C        PRINT THE NAME AND THE VALUE OF VARIABLES
C
         WRITE (6.1) NAME,X
         RETURN
    1    FORMAT (30X,A6,1H=,F15.2)
         END
```

A7 Table A7.1 Optimum Dimensions of Compressed Struts of Welded Square Box Section

Design method described in Section 3.2. Computer method: backtrack programming.
Material: steel 37, $R_y = 240$ MPa, $R_u = 190$ MPa, $E = 210$ GPa. N in [kN],
L in [m], b and t in [mm], A in [mm²]

N	L	$b \times t$	A	L	$b \times t$	A	L	$b \times t$	A
100	2.4	80 × 3	960	3.2	100 × 3	1200	4.0	120 × 3	1440
	2.6	90 × 3	1080	3.4	100 × 3	1200	4.2	120 × 3	1440
	2.8	90 × 3	1080	3.6	110 × 3	1320	4.4	130 × 3	1560
	3.0	100 × 3	1200	3.8	110 × 3	1320	4.6	130 × 3	1560
125	2.4	90 × 3	1080	3.2	100 × 3	1200	4.0	120 × 3	1440
	2.6	90 × 3	1080	3.4	110 × 3	1320	4.2	130 × 3	1560
	2.8	100 × 3	1200	3.6	110 × 3	1320	4.4	130 × 3	1560
	3.0	100 × 3	1200	3.8	120 × 3	1440	4.6	130 × 3	1560
150	2.4	100 × 3	1200	3.2	110 × 3	1320	4.0	130 × 3	1560
	2.6	100 × 3	1200	3.4	110 × 3	1320	4.2	130 × 3	1560
	2.8	100 × 3	1200	3.6	120 × 3	1440	4.4	130 × 3	1560
	3.0	110 × 3	1320	3.8	120 × 3	1440	4.6	140 × 3	1680
175	2.4	100 × 3	1200	3.2	120 × 3	1440	4.0	130 × 3	1560
	2.6	110 × 3	1320	3.4	120 × 3	1440	4.2	130 × 3	1560
	2.8	110 × 3	1320	3.6	120 × 3	1440	4.4	140 × 3	1680
	3.0	110 × 3	1320	3.8	130 × 3	1560	4.6	140 × 3	1680
200	2.4	110 × 3	1320	3.2	120 × 3	1440	4.0	140 × 3	1680
	2.6	120 × 3	1440	3.4	130 × 3	1560	4.2	140 × 3	1680
	2.8	120 × 3	1440	3.6	130 × 3	1560	4.4	140 × 3	1680
	3.0	120 × 3	1440	3.8	130 × 3	1560	4.6	150 × 3	1800
225	2.4	100 × 4	1600	3.2	120 × 4	1920	4.0	130 × 4	2080
	2.6	110 × 4	1760	3.4	120 × 4	1920	4.2	140 × 4	2240
	2.8	110 × 4	1760	3.6	120 × 4	1920	4.4	140 × 4	2240
	3.0	110 × 4	1760	3.8	130 × 4	2080	4.6	150 × 4	2400
250	2.4	110 × 4	1760	3.2	120 × 4	1920	4.0	140 × 4	2240
	2.6	110 × 4	1760	3.4	130 × 4	2080	4.2	140 × 4	2240
	2.8	120 × 4	1920	3.6	130 × 4	2080	4.4	150 × 4	2400
	3.0	120 × 4	1920	3.8	130 × 4	2080	4.6	150 × 4	2400
275	2.4	120 × 4	1920	3.2	130 × 4	2080	4.0	140 × 4	2240
	2.6	120 × 4	1920	3.4	130 × 4	2080	4.2	150 × 4	2400
	2.8	120 × 4	1920	3.6	130 × 4	2080	4.4	150 × 4	2400
	3.0	120 × 4	1920	3.8	140 × 4	2240	4.6	150 × 4	2400

Table A7.1 (cont.)

N	L	b × t	A	L	b × t	A	L	b × t	A
	2.4	120 × 4	1920	3.2	130 × 4	2080	4.0	150 × 4	2400
300	2.6	130 × 4	2080	3.4	140 × 4	2240	4.2	150 × 4	2400
	2.8	130 × 4	2080	3.6	140 × 4	2240	4.4	150 × 4	2400
	3.0	130 × 4	2080	3.8	140 × 4	2240	4.6	160 × 4	2560
	2.4	130 × 4	2080	3.2	140 × 4	2240	4.0	150 × 4	2400
325	2.6	130 × 4	2080	3.4	140 × 4	2240	4.2	160 × 4	2560
	2.8	140 × 4	2240	3.6	150 × 4	2400	4.4	160 × 4	2560
	3.0	140 × 4	2240	3.8	150 × 4	2400	4.6	160 × 4	2560
	2.4	140 × 4	2240	3.2	150 × 4	2400	4.0	160 × 4	2560
350	2.6	140 × 4	2240	3.4	150 × 4	2400	4.2	160 × 4	2560
	2.8	140 × 4	2240	3.6	150 × 4	2400	4.4	160 × 4	2560
	3.0	·140 × 4	2240	3.8	160 × 4	2560	4.6	170 × 4	2720
	2.4	150 × 4	2400	3.2	150 × 4	2400	4.0	160 × 4	2560
375	2.6	150 × 4	2400	3.4	160 × 4	2560	4.2	170 × 4	2720
	2.8	150 × 4	2400	3.6	160 × 4	2560	4.4	170 × 4	2720
	3.0	150 × 4	2400	3.8	150 × 5	3000	4.6	170 × 4	2720
	2.4	150 × 4	2400	3.2	140 × 5	2800	4.0	150 × 5	3000
400	2.6	160 × 4	2560	3.4	140 × 5	2800	4.2	160 × 5	3200
	2.8	160 × 4	2560	3.6	150 × 5	3000	4.4	160 × 5	3200
	3.0	140 × 5	2800	3.8	150 × 5	3000	4.6	160 × 5	3200
	2.4	160 × 4	2560	3.2	150 × 5	3000	4.0	160 × 5	3200
425	2.6	140 × 5	2800	3.4	150 × 5	3000	4.2	160 × 5	3200
	2.8	140 × 5	2800	3.6	150 × 5	3000	4.4	160 × 5	3200
	3.0	140 × 5	2800	3.8	150 × 5	3000	4.6	170 × 5	3400
	2.4	140 × 5	2800	3.2	150 × 5	3000	4.0	160 × 5	3200
450	2.6	140 × 5	2800	3.4	150 × 5	3000	4.2	170 × 5	3400
	2.8	150 × 5	3000	3.6	160 × 5	3200	4.4	170 × 5	3400
	3.0	150 × 5	3000	3.8	160 × 5	3200	4.6	170 × 5	3400
	2.4	150 × 5	3000	3.2	160 × 5	3200	4.0	170 × 5	3400
475	2.6	150 × 5	3000	3.4	160 × 5	3200	4.2	170 × 5	3400
	2.8	150 × 5	3000	3.6	160 × 5	3200	4.4	170 × 5	3400
	3.0	150 × 5	3000	3.8	160 × 5	3200	4.6	180 × 5	3600
	2.4	150 × 5	3000	3.2	160 × 5	3200	4.0	170 × 5	3400
500	2.6	160 × 5	3200	3.4	160 × 5	3200	4.2	180 × 5	3600
	2.8	160 × 5	3200	3.6	170 × 5	3400	4.4	180 × 5	3600
	3.0	160 × 5	3200	3.8	170 × 5	3400	4.6	180 × 5	3600

Table A 7.1 (concl.)

N	L	b × t	A	L	b × t	A	L	b × t	A
525	2.4	160 × 5	3200	3.2	170 × 5	3400	4.0	180 × 5	3600
	2.6	160 × 5	3200	3.4	170 × 5	3400	4.2	180 × 5	3600
	2.8	160 × 5	3200	3.6	170 × 5	3400	4.4	180 × 5	3600
	3.0	170 × 5	3400	3.8	180 × 5	3600	4.6	190 × 5	3800
550	2.4	140 × 6	3360	3.2	170 × 5	3400	4.0	180 × 5	3600
	2.6	170 × 5	3400	3.4	180 × 5	3600	4.2	190 × 5	3800
	2.8	170 × 5	3400	3.6	180 × 5	3600	4.4	190 × 5	3800
	3.0	170 × 5	3400	3.8	180 × 5	3600	4.6	190 × 5	3800
575	2.4	170 × 5	3400	3.2	180 × 5	3600	4.0	190 × 5	3800
	2.6	170 × 5	3400	3.4	180 × 5	3600	4.2	190 × 5	3800
	2.8	150 × 6	3600	3.6	180 × 5	3600	4.4	190 × 5	3800
	3.0	180 × 5	3600	3.8	190 × 5	3800	4.6	200 × 5	4000
600	2.4	150 × 6	3600	3.2	190 × 5	3800	4.0	190 × 5	3800
	2.6	180 × 5	3600	3.4	190 × 5	3800	4.2	180 × 6	4320
	2.8	180 × 5	3600	3.6	190 × 5	3800	4.4	200 × 5	4000
	3.0	180 × 5	3600	3.8	170 × 6	4080	4.6	180 × 6	4320
625	2.4	180 × 5	3600	3.2	190 × 5	3800	4.0	180 × 6	4320
	2.6	190 × 5	3800	3.4	190 × 5	3800	4.2	180 × 6	4320
	2.8	190 × 5	3800	3.6	170 × 6	4080	4.4	180 × 6	4320
	3.0	190 × 5	3800	3.8	180 × 6	4320	4.6	190 × 6	4560
650	2.4	190 × 5	3800	3.2	170 × 6	4080	4.0	180 × 6	4320
	2.6	190 × 5	3800	3.4	180 × 6	4320	4.2	190 × 6	4560
	2.8	190 × 5	3800	3.6	180 × 6	4320	4.4	190 × 6	4560
	3.0	200 × 5	4000	3.8	180 × 6	4320	4.6	190 × 6	4560
675	2.4	200 × 5	4000	3.2	180 × 6	4320	4.0	190 × 6	4560
	2.6	200 × 5	4000	3.4	180 × 6	4320	4.2	190 × 6	4560
	2.8	200 × 5	4000	3.6	180 × 6	4320	4.4	190 × 6	4560
	3.0	180 × 6	4320	3.8	180 × 6	4320	4.6	200 × 6	4800
700	2.4	200 × 5	4000	3.2	180 × 6	4320	4.0	190 × 6	4560
	2.6	210 × 5	4200	3.4	180 × 6	4320	4.2	190 × 6	4560
	2.8	210 × 5	4200	3.6	190 × 6	4560	4.4	200 × 6	4800
	3.0	180 × 6	4320	3.8	190 × 6	4560	4.6	200 × 6	4800
725	2.4	210 × 5	4200	3.2	190 × 6	4560	4.0	200 × 6	4800
	2.6	180 × 6	4320	3.4	190 × 6	4560	4.2	200 × 6	4800
	2.8	180 × 6	4320	3.6	190 × 6	4560	4.4	200 × 6	4800
	3.0	190 × 6	4560	3.8	190 × 6	4560	4.6	200 × 6	4800

References and Bibliography

Agha, G. T. and Nelson, R. B. (1976) Method for the optimum design of truss-type structures. *AIAA J.*, 14, 436–445.

Akita, Y. and Kitamura, K. (1972) Studies on nonlinear programming optimization for ship structures and optimum design of box and I-section girders. *Japan Shipbuilding and Marine Engineering*, 6, No. 5. 15–27.

Allen, H. G. (1966) Optimum design of sandwich struts and beams. *Plastic in Building Structures Conference.* London, 1965, Oxford, Pergamon, 201–209.

Allen, H. G. (1973) Sandwich panels with thick or flexurally stiff faces. *Sheet Steel in Building. Papers of a Meeting, London, 1972,* London, Iron and Steel Inst. 10–18.

Anderson, D. and Islam, M. A. (1979) Design of multistorey frames to sway deflection limitations. *Struct. Eng.,* 57B, 11–17.

Annamalai, N. (1970) Cost optimization of welded plate girders. Dissertation, Purdue Univ. Indianapolis, Ind.

Annamalai, N., Lewis, A. D. M. and Goldberg, J. E. (1972) Cost optimization of welded plate girders. *J. Struct. Div. Proc. ASCE,* 98, 2235–2246.

Anraku, H. (1976) Optimum design of steel frame subjected to dynamic earthquake forces. *Proc. 10th IABSE Congress, Tokyo. Preliminary Report,* Zürich, 93–98.

Azad, A. K. (1980) Continuous steel I-girders: optimum proportioning. *J. Struct. Div. Proc. ASCE,* 106, 1543–1555.

Bahke, E. (1970) Neue Materialflußgestaltung im Güterverkehr. *Fördern und Heben,* 20, 783–790.

Banichuk, N. V. (1980) *Optimizatsiya form uprugikh tel.* Moskva, Nauka.

Banichuk, N. V. (1982) Up-to-date problems of the optimization of structures (in Russian). *Izvestiya Vuzov Mekhanika Tverdogo Tela,* 110–124.

Barta, J. (1957) On the minimum weight of certain redundant structures. *Acta Techn. Hung.,* 18, 67–75.

Batishchev, D. I. (1975) *Poiskovye metody optimal'nogo proektirovaniya.* Moskva, Sovetskoe Radio.

Beer, H. and Schulz, G. (1970) Bases théoriques des courbes européennes de flambement. *Constr. Métall.,* 7, No. 3, 37–57.

Beleňa, J. I. and Neždanov, K. K. (1979) Experimental investigation of the fatigue behaviour of the compression zone of crane runway girder webs (in Slovak). *Stavebnicky Časopis,* 27, 441–456.

Bellagamba, L. and Yang, T. Y. (1981) Minimum-mass truss structures with constraints on fundamental natural frequency. *AIAA J.,* 19, 1452–1458.

Bennett, J. A. (1980) Topological structural synthesis. *Computers & Structures,* 12, 273–280.

Bigelow, R. H. and Gaylord, E. H. (1967) Design of steel frames for minimum weight. *J. Struct. Div. Proc. ASCE,* 93, 109–131.

Bitner, J. R. and Reingold, E. M. (1975) Backtrack programming techniques. *Communications of ACM*, **18**, 651–656.

Bo, G. M., Capurro, P. M. and Daddi, I. (1974) Sulla razionalizzazione dei profili ad I per flessione e taglio. *Costruz. Metall.*, **26**, No. 2, 102–112.

Boguslavskiy, P. E. (1961) *Metallicheskie konstruktsii gruzopodyemnykh mashin i sooruzheniy*. Moskva, Mashgiz.

Box, M. J. (1965) A new method of constrained optimization and a comparison with other methods. *Computer J.*, **8**, 42–52.

Braham, M., Grimault, J. P. Massonnet, C., Mouty, J. and Rondal, J. (1980) Buckling of thin-walled hollow sections. Cases of axially-loaded rectangular sections. *Acier-Stahl-Steel*, **45**, 30–36.

Brandt, A. M. ed. (1977) *Kryteria i metody optymalizacji konstrukcji*. Warszawa, Panstwowe Wyd. Naukowe.

Bronowicki, A. J. and Felton, L. P. (1975) Optimum design of continuous thin-walled beams. *Int. J. Numer. Methods in Engineering*, **9**, 711–720.

Brown, D. M. and Ang, A. H. (1966) Structural optimization by nonlinear programming. *J. Struct. Div. Proc. ASCE*, **92**, 319–340.

Brozzetti, J. and Lescouarc'h, Y. (1973) Prédimensionnement des structures métalliques à barres par la méthode statique et la programmation linéaire. *Revue Française de Mécanique*, No. 48. 5–14.

Bryzgalin, G. I. (1981) Optimization with more objective functions in the case of a composite plate (in Russian). *Mekhanika Kompozitnykh Materialov*, No. 1. 70–76.

Caldwell, J. B. and Hewitt, A. D. (1976) Cost effective design of ship structures. *Metal Construction*, **8**, No. 2, 64–67.

Carroll, C. W. (1961) The created response surface technique for optimizing nonlinear, restrained systems. *Operations Research*, No. 2, 169–184.

Carskaddan, P. S. (1968) Shear buckling of unstiffened hybrid beams. *J. Struct. Div. Proc. ASCE*, **94**, 1965–1990.

Cassis, J. H. (1974) Optimum design of structures subjected to dynamic loads. *Ph. D. Thesis*, Univ. of California, Los Angeles.

Cassis, J. H. and Schmit, L. A. jr. (1976a) On implementation of the extended interior penalty function. *Int. J. Numer. Methods in Eng.*, **10**, 3–23.

Cassis, J. H. and Schmit, L. A. jr. (1976b) Optimum structural design with dynamic constraints. *J. Struct. Div. Proc. ASCE*, **102**, 2053–2071.

Cella, A. and Logcher, R. D. (1971) Automated optimum design from discrete components. *J. Struct. Div. Proc. ASCE*, **97**, 175–189.

Cheng, F. Y. and Botkin, M. E. (1976) Nonlinear optimum design of dynamic damped frames. *J. Struct. Div. Proc. ASCE*, **102**, 609–627.

Chen, W. F. and Atsuta, T. (1976–77) *Theory of beam-columns*. Vols 1–2. New York, McGraw Hill.

Chizhas, A. P. and Nagyavichyus, Yu. A. (1978) Optimum design of rod structures subjected to bending . . . (in Russian). *Litovskiy Mekhanicheskiy Sbornik*. Vilnyus. No. 19. 63–72.

Cho, H. N. (1972) A discrete formulation of minimum cost design of framed structures. *Dissertation*. Michigan State Univ.

Chong, K. P. (1976) Optimization of unstiffened hybrid beams. *J. Struct. Div. Proc. ASCE*, **102**, 401–409.

Cinquini, C. and Sacchi, G. (1980) Problems of optimal design for elastic and plastic structures. *J. méc. appl.*, **4**, 31–59.

Cohn, M. Z. and Maier, G. eds (1977) *Engineering plasticity by mathematical programming*. New York, Pergamon.

Crawford, A. B. and Jenkins, W. M. (1980) Optimum design of steel roof structures. *Struct. Eng.*, **58A**, 317–325.

Czesany, G. (1972) *Kostenrechnen beim Schweissen.* Essen, Vulkan-Verlag.

Dafalias, Y. F. and Dupuis, G. (1972) Minimum weight design of continuous beams under displacements and stress constraints. *J. Optimiz. Theory and Appl.,* 9, 137–154.

Davidon, W. C. (1959) *Variable metric method for minimization.* Argonne Nat. Lab. Illinois.

Davidson, J. W., Felton, L. P. and Hart, G. C. (1980) On reliability-based structural optimization for earthquakes. *Computers & Structures,* 12, 99–105.

Dekhtyar', A. S. and Rasskazov, A. O. (1980) Optimum design of sandwich plates (in Russian). *Problemy Prochnosti,* No. 2, 77–80.

Delesques, R. (1979) Résistance des âmes de poutres sans raidisseurs intermédiaires. *Constr. Métall.,* 11, No. 2, 5–19.

Demokritov, V. N. (1962) Design of the main girders of overhead travelling cranes (in Russian). *Vestnik Mashinostroeniya,* 42, No. 4, 30–33.

Design of hybrid steel beams. (1968) *J. Struct. Div. Proc. ASCE,* 94, 1397–1426.

Design recommendations for hollow section joints. (1982) *Welding in the World,* 20, No.3/4, 53–78.

Design rules for arc-welded connexions in steel submitted to static loads. (1976) *Welding in the World,* 14, No. 5/6, 132–149.

De Silva, B. M. and Grant, G. N. (1973) Comparison of some penalty function based optimization procedures for synthesis of a planar truss. *Int. J. Numer. Methods in Engng,* 7, 155–173.

Dobbs, M. W. and Nelson, R. B. (1976) Application of optimality criteria to automated structural design. *AIAA J.,* 14, 1436–1443.

Dobbs, M. W. and Nelson, R. B. (1978) Minimum weight design of stiffened panels with fracture constraints. *Computers & Structures,* 8, 753–759.

Douwen, A. A. van (1981) Design for economy in bolted and welded connections. *Joints in Structural Steelwork.* London – Plymouth, Pentech Press. 5.18–5.35.

Dupuis, G. (1971) Optimal design of statically determinate beams subject to displacement and stress constraints. *AIAA J.,* 9, 981–984.

Dwight, J. B. (1975) Adaption of Perry formula to represent the new European steel column-curves. *Steel Construction* (Australia), 9, No. 1, 11–23.

Eschenauer, H. (1981) Anwendung der Vektoroptimierung bei räumlichen Tragstrukturen. *Stahlbau,* 50, 110–115.

European recommendations for steel construction. (1978) European Convention for Constructional Steelwork.

Farkas, J. (1963) Effect of residual welding stresses on buckling of compressed struts (in Hungarian). *Mélyépítéstudományi Szemle,* 13, 474–478.

Farkas, J. (1965) Design of simply supported square plates stiffened on one side . . . *Proc. 2nd Conf. Dimensioning,* Budapest, Akadémiai Kiadó, 65–77.

Farkas, J. (1966a) Methods of control of distortions of prismatic bars due to longitudinal welds (in Hungarian). *Mélyépítéstudományi Szemle,* 16, 139–145.

Farkas, J. (1966b) Vergleichende Auswertung von Berechnungsmethoden . . . *Referaty III. Konf. Konstrukcje Metalowe,* Warszawa, PZITB, T. 1. 117–128.

Farkas, J. (1967) Optimum design of crane bridge and runway girders of box cross-section (in Hungarian). *Proc. 5th Conf. Materials Handling,* Budapest. MTESZ-KAB, 17/1 – – 17/16.

Farkas, J. (1968) Berechnungsprobleme von einseitig in zwei Richtungen versteiften Quadratplatten. *Wiss. T. TH Dresden,* 17, 385–388.

Farkas, J. (1969a) Festigkeitseigenschaften von geschweißten, auf Biegung optimal bemessenen I- und Kastenträgern. *Acta Techn. Hung.,* 66, 427–439.

Farkas, J. (1969b) Analytische Betrachtung von Methoden zur Verminderung der Krümmung infolge Längsschrumpfung. *Schweisstechnik DDR,* 19, 406–408.

Farkas, J. (1970a) Geschweisste Zellenplatten. *Vorträge der 6. Konferenz für Schweisstechnik,* Budapest. Wiss. Verein für Maschinenbau, 103–108.

Farkas, J. (1970b) Minimum volume design of a welded sandwich square plate . . . *Referaty IV. Konf. Konstrukcji Metalowe,* Warszawa, T. 2. 57–62.

Farkas, J. (1972) Optimalbemessung und Vergleich von biegebeanspruchten dünnwandigen Trägern mit Kasten-, Kreisrohr- und Ovalquerschnitt. *Acta Techn. Hung.,* **72,** 377––388.

Farkas, J. (1973a) Minimum volume design of welded stiffened plates . . . *Proc. 4th Conf. Dimensioning, Budapest,* 1971. Akadémiai Kiadó, 313–324.

Farkas, J. (1973b) Schwingungsdämpfung von Schweisskonstruktionen. *Schweisstechnik DDR,* **23,** 508–512.

Farkas, J. (1974a) *Metal structures.* University textbook in Hungarian. Budapest, Tankönyvkiadó.

Farkas, J. (1974b) Structural synthesis of press frames having columns and cross beams of welded box cross-section. *Acta Techn. Hung.,* **79,** 191–201.

Farkas, J. (1974c) Static investigations and optimum design of welded cellular plates (in Hungarian). *Gép,* **26,** 233–238.

Farkas, J. (1976) Structural synthesis of welded cell-type plates. *Acta Techn. Hung.,* **83,** 117–131.

Farkas, J. and Tímár, I. (1976) Vibration damping of sandwich beams (in Hungarian). *Járművek, Mezőgazdasági Gépek,* **23,** 207–211.

Farkas, J. (1977a) Kostenoptimieren von geschweissten Zellenplatten. *Berichte VII. Int Kongress über Anwendungen der Mathematik in den Ingenieurwissenschaften,* Weimar, 1975. Berlin, VEB Verlag für Bauwesen. 163–167.

Farkas, J. (1977b) Statische Biegesteifigkeit und Schwingungsdämpfung von Sandwichbalken. *Proc. Conf. Materials Handling,* Varna, 1976. Sofia, 106–122.

Farkas, J. (1977c) The effect of residual welding stresses on the buckling strength of compressed plates. *Proc. Regional Colloquium on Stability of Steel Structures,* Budapest, 299–306.

Farkas, J. (1977d) Optimum design of welded structures (in Hungarian). *Gép,* **29,** 112–115.

Farkas, J. (1977e) Optimum design of compressed columns of constant welded square box cross-section considering the effect of residual welding stresses. *Acta Techn. Hung.,* **84,** 335–348.

Farkas, J. (1977f) Optimum design of metal structures. *Dissertation* for academic degree D. Sc. (in Hungarian).

Farkas, J. and Tímár, I. (1977) Structural optimization by means of the SUMT nonlinear programming method (in Hungarian). *Járművek, Mezőgazd. Gépek,* **24,** 103–108.

Farkas, J. (1978a) Economy of welded unstiffened I-section rods. *Regional Coll. Stability of Steel Structures,* Budapest, 1977, Final Report, Budapest. 107–112.

Farkas, J. (1978b) Elastic and plastic minimum weight design of the welded I-sections of a single-bay pitched-roof portal frame (in Hungarian). *Magyar Építőipar,* **27,** 490–497.

Farkas, J. (1978c) Minimization of the cross-section area of welded unstiffened plate and box girders subjected to bending and shear. *Acta Techn. Hung.,* **87,** 295–306.

Farkas, J. (1979a) Economy of welded I-sections (in Hungarian). *Gép,* **31,** 113–116.

Farkas, J. (1979b) Minimum weight design of statically indeterminate rod structures . . . *Proc. 6th Int. Conf. Metal Constructions,* Katowice, Vol. 1. 57–68.

Farkas, J. (1980a) Optimum design of metal structures by backtrack programming. *11th IABSE Congress Vienna. Final Report,* Zürich. 597–602.

Farkas, J. (1980b) Optimale Dimensionierung von Kranhauptträgern mit Kastenprofil. *"Kran '80" III. Koll. über Krane. Vorträge.* Budapest, Wiss. Verein für Maschinenbau. 29–41.

Farkas, J. (1980c) Optimum design for bending and ultimate shear strength of hybrid I-beams. *Acta Techn. Hung.,* **90,** 259–273.

Farkas, J. and Szabó, L. (1980) Optimum design of beams and frames of welded I-sections by means of backtrack programming. *Acta Techn. Hung.,* **91,** 121–135.

Farkas, J. (1981) Anwendung der Backtrack-Programmierungsmethode auf die Optimierung von geschweissten Stabtragwerken. *Berichte IX. Int. Kongress über Anwendung der Mathematik in den Ingenieurwissenschaften,* Weimar. Heft 1. 70–73.

Farkas, J. (1982a) Optimal design of steel trusses and frames – a survey of selected literature. *Publ. Techn. Univ. Heavy Ind. Ser. C. Machinery,* 36, 209–226.

Farkas, J. (1982b) Minimum cost design of welded square cellular plates. *Publ. Techn. Univ. Heavy Ind. Ser. C. Machinery,* 37, 111–130.

Farkas, J. and Jármai, K. (1982) Structural synthesis of sandwich beams with outer layers of box-section. *J. Sound and Vibr.,* 84, 47–56.

Farkas, J. (1983) *Metal Structures.* Univ. textbook in Hungarian. 2nd ed. Budapest, Tankönyvkiadó.

Faulkner, D., Adamczak, J. C., Snyder, G. J. and Vetter, M. F. (1973) Synthesis of welded grillages to withstand compression and normal loads. *Computers & Structures,* 3, 221–246.

Felton, L. P. and Dobbs, M. W. (1967) Optimum design of tubes for bending and torsion. *J. Struct. Div. Proc. ASCE,* 93, 185–200.

Felton, L. P. and Hofmeister, L. D. (1968) Optimized components in truss synthesis. *AIAA J.,* 6, 2434–2435.

Felton, L. P. and Nelson, R. B. (1971) Optimized components in frame synthesis. *AIAA J.,* 9, 1027–1031.

Felton, L. P. and Fuchs, M. B. (1981) Simplified direct optimization of tubular truss structures. *Int. J. Numer. Methods in Eng.,* 17, 601–613.

Feng, T. T., Arora, J. S. and Haug, E. J. jr. (1977) Optimal structural design under dynamic loads. *Int. J. Numer. Methods in Eng.* 11, 39–52.

Ferjenčik, R. and Tocháček, M. (1980) Economic design of prestressed steel web-plate girders in elastic and elasto-plastic states (in Slovak). *Stavebnický Časopis,* 28, 169–198.

Fiacco, A. V. and McCormick, G. P. (1968) *Nonlinear programming: sequential unconstrained minimization techniques.* New York, Wiley.

Fletcher, R. and Powell, M. J. D. (1963) A rapidly convergent descent method for minimization. *Computer J.,* 6, 163–168.

Fletcher, R. (1980–81) *Practical methods of optimization.* Vols 1–2. Chichester, Wiley & Sons.

Fleury, C. (1979) A unified approach to structural weight minimization. *Computer Methods in Appl. Mech. and Eng.,* 20, 17–38.

Fleury, C. (1980) Optimization of large flexural finite element systems. *Collect. Publ. Univ. Liège, Fac. Sci. Appl.* No. 84. 29–42.

Fleury, C. and Schmit, L. A. (1980) Primal and dual methods in structural optimization. *J. Struct. Div. Proc. ASCE,* 106, 1117–1133.

Fox, R. L. and Schmit, L. A. (1966) Advances in the integrated approach to structural synthesis. *J. Spacecraft,* 3, 858–866.

Fox, R. L. and Kapoor, M. P. (1970) Structural optimization in the dynamics response regime: a computational approach. *AIAA J.,* 8, 1798–1804.

Fox, R. L. (1971) *Optimization methods for engineering design.* Reading–London, Addison–Wesley.

Frieze, P. A., Dowling, P. J. and Hobbs, R. E. (1977) Ultimate load behaviour of plates in compression. *Steel Plated Structures.* London, Crosby Lockwood Staples, 24–50.

Frieze, P. A. (1980) Behaviour and design of thin-walled rectangular hollow beams. *Thin-walled Structures.* London, Granada, 455–477.

Frind, E. O. and Wright, P. M. (1975) Gradient methods in optimum structural design. *J. Struct. Div. Proc. ASCE,* 101, 939–956.

Fukumoto, Y. and Ito, M. (1981) Minimum weight plastic design of continuous beams. *J. Struct. Div. Proc. ASCE*, **107**, 1263–1277.

Funaro, J. R. (1974) Optimum structural design with deflection constraints. *Dissertation*. Columbia Univ.

Gallagher, R. H. and Zienkiewicz, O. C. eds (1973) *Optimum structural design*. New York, Wiley.

Geiger, M. (1974) Beitrag zur rechnerunterstützten Auslegung von Pressengestellen. Essen, Girardet.

Gellatly, R. A., Helenbrook, R. G. and Kocher, L. H. (1978) Multiple constraints in structural optimization. *Int. J. Numer. Methods in Eng.*, **13**, 297–309.

Gellatly, R. A. and Thom, R. D. (1980) *Force method optimization*. US Air Force Wright Aeronautical Lab. AFWAL-TR-80-3006.

Gerasimov, E. N. and Repko, V. N. (1978) Optimization with more objective functions (in Russian). *Prikladnaya Mekhanika*, **14**, No. 11, 72–78.

Glankwahmdee, A., Liebman, J. S. and Hogg, G. L. (1979) Unconstrained discrete nonlinear programming. *Engineering Optimization*, **4**, 95–107.

Glushkov, G., Yegorov, I. and Yermolov, V. (1975) *Formulas for designing frames*. Moscow, Mir.

Gol'dstein, Yu. B. and Solomeshch, M. A. (1980) Variatsionnye zadachi statiki optimal'-nykh strezhnevykh sistem. Leningrad, LGU.

Golomb, S. W. and Baumert, L. D. (1965) Backtrack programming. *J. Assoc. Computing Machinery*, **12**, 516–524.

Golubenko, V. V., Skirko, V. F. and Seletskiy, V. F. (1975) Belt conveyor bridges of circular hollow cross-section (in Russian). *Promyshlennoe Stroitel'stvo*, No. 2, 40–41.

Grierson, D. E. and Schmit, L. A. jr. (1982) Synthesis under service and ultimate performance constraints. *Computers & Structures*, **15**, 405–417.

Grosskopf, P. and Winkler, Th. (1973) Auslegung von GFK-Hartschaum-Verbundwerkstoffen. *Kunststoffe*, **63**, 881–888.

Haftka, R. T. and Starnes, J. H. (1975) Applications of a quadratic extended interior penalty function for structural optimization. *AIAA Paper*, No. 764.

Haftka, R. T. and Prasad, B. (1981) Optimum structural design with plate bending elements – a survey. *AIAA J.*, **19**, 517–522.

Hansen, H. R. (1974) Application of optimization methods within structural design. *Computers & Structures*, **4**, 213–220.

Harless, R. I. (1980) A method for synthesis of optimal weight structures. *Computers & Structures*, **12**, 791–804.

Haug, E. J. and Arora, J. S. (1978) *Applied optimal design*. New York, Wiley-Interscience.

Haug, E. J. and Rousselet, B. (1980) Design sensitivity analysis in structural mechanics. 1. Static response variations. *J. Struct. Mech.*, **8**, 17–41.

Herden, G. (1967) *Schweiß- und Schneidtechnologie*. 2. Aufl. Berlin, VEB Verlag Technik.

Himmelblau, D. M. (1971) *Applied nonlinear programming*. New York, McGraw-Hill.

Holst, O. (1974) *Automatic design of plane frames*. Techn. Univ. of Denmark, Lyngby, Struct. Res. Lab. Report No. R 53.

Holt, E. C. and Heithecker, G. L. (1969) Minimum weight proportions for steel girders. *J. Struct. Div. Proc. ASCE*, **95**, 2205–2217.

Horne, M. R. (1979) *Plastic theory of structures*. 2nd ed. Oxford, Pergamon Press.

Horne, M. R. and Morris, L. I. (1981) *Plastic design of low rise frames*. London Granada.

Hupfer, P. (1970) *Optimierung von Baukonstruktionen*. Stuttgart, Teubner.

Hupfer, P. (1974) Entwurf des Tischkörpers von Zweiständerpressen. *Maschinenbautechnik*, **23**, 210–214.

Icerman, L. J. (1969) Optimal structural design for given dynamic deflection. *Int. J. Solids and Struct.*, **5**, 473–490.

Imai, K. and Schmit, L. A. jr. (1981) Configuration optimization of trusses. *J. Struct. Div. Proc. ASCE*, **107**, 745–756.

Imai, K. and Shoji, M. (1981) Minimum cost design of framed structures by the mini-max dual method. *Int. J. Numer. Methods in Eng.* **17**, 213–229.

Isakson, G., Pardo, H. and Lerner, E. (1978) ASOP-3: A program for optimum structural design to satisfy strength and deflection constraints. *J. Aircraft*, **15**, 422–428.

Isreb, M. (1978) DESAP-1: A structural synthesis with stress and local instability constraints. *Computers & Structures*, **8**, 243–256.

Iyengar, N. G. R. and Gupta, S. K. (1980) *Programming methods in structural design*. New Delhi – Madras, East–West Press.

Izbrannye zadachi po stroitel'noy mekhanike i teorii uprugosti. Red. Abovskiy, N. P. Moskva, Stroyizdat, 1978. Glava 4. Optimal'noe proektirovanie konstruktsiy. 82–187.

Jármai, K. (1982) Optimal design of welded frames by complex programming method. *Publ. Techn. Univ. Heavy Ind. Ser. C. Machinery*, **37**, 79–95.

Jesenský, M. (1973) *Navrhovanie, výpočet a zváranie rámov tvárniacich strojov.* Bratislava, Výskumný Ústav Zváračský.

Johnson, R. C. (1980) *Optimum design of mechanical elements.* 2nd ed. New York, Wiley.

Johnston, B. G. ed. (1976) *Guide to stability design criteria for metal structures.* 3rd ed. New York, Wiley.

Juhás, P. (1978) Optimum design of steel I-beams considering the influence of web buckling (in Slovak). *Stavebnický Časopis*, **26**, 639–654.

Kaliszky, S. (1975) *Theory of plasticity* (in Hungarian). Budapest, Akadémiai Kiadó.

Kartvelishvili, V. M., Mironov, A. A. and Samsonov, A. M. (1981) Numerical method for the solution of optimization problems of stiffened structures (in Russian). *Izv. AN SSSR Mekhanika tverdogo Tela,* No. 2, 93–103.

Kato, B. and Okumura, T. (1976) Structural behaviour including hybrid construction. *10th IABSE Congress Tokyo. Introductory Report,* Zürich. 199–223.

Kavlie, D., Kowalik, J. and Moe, J. (1966) *Structural optimization by means of nonlinear programming.* The Techn. Univ. of Norway, Dept. of Ship Struct. Trondheim.

Kavlie, D. and Moe, J. (1969) Application of nonlinear programming to optimum grillage design with nonconvex sets of variables. *Int. J. Numer. Methods in Eng.* **1**, 351–378.

Kavlie, D. (1970) Optimum design of statically indeterminate structures. *Dissertation.* Univ. of California, Berkeley.

Kavlie, D. and Moe, J. (1971) Automated design of frame structures. *J. Struct. Div. Proc. ASCE,* **97**, 33–62.

Khalifa, M. M. K. and Merwin, J. E. (1977) Optimum plastic design of space frames. *Proc. Inst. Civ. Engrs Part 2.,* **63**, 769–783.

Khan, M. R., Thornton, W. A. and Willmert, K. D. (1978) Optimality criterion techniques applied to mechanical design. *Trans. ASME J. Mech. Design,* **100**, 319–327.

Khan, M. R., Willmert, K. D. and Thornton, W. A. (1979) An optimality criterion method for large-scale structures. *AIAA J.,* **17**, 753–761.

Khan, M. R. and Willmert, K. D. (1981) An efficient optimality criterion method for natural frequency constrained structures. *Computers & Structures,* **14**, 501–507.

Kholopov, I. S. and Loseva, I. V. (1980) Optimization of cross-sections of I-beam elements of steel frames based on a discrete model (in Russian). *Izv. Vuzov Stroitel̆ stvo i Arkhitektura*, No. 1, 54–58.

Khot, N. S., Berke, L. and Venkayya, V. B. (1979) Comparison of optimality criteria algorithms for minimum weight design of structures. *AIAA J.*, 17, 182–190.

Khot, N. S. (1981) Algorithms based on optimality criteria to design minimum weight structures. *Engineering Optimization*, 5, 73–90.

Kicher, T. P. (1966) Optimum design – minimum weight versus fully stressed. *J. Struct. Div. Proc. ASCE*, 92, 265–280.

Kilp, K. H. (1970) Untersuchungen an Pressengestellen bei statischer und dynamischer Belastung. *Maschinenbautechnik*, 19, 114–126.

Kirsch, U. and Benardout, D. (1980) Optimal design of elastic trusses by approximate equilibrium. *Computer Methods in Appl. Mech. and Eng.*, 22, 347–359.

Kirsch, U. (1982) *Optimum structural design*. New York, McGraw-Hill.

Kiusalaas, J. (1973) Optimal design of structures with buckling constraints. *Int. J. Solids and Struct.*, 9, 863–878.

Klinnert, M. and Möbius, W. (1975) Festigkeits- und Schwingungsverhalten der Kreuzwagenscheidemaschine ZIS 806. *ZIS Mitt.*, 17, 1020–1024.

Klinov, S. I. (1973) Optimum parameters of hybrid beams (in Russian). *Sbornik Trudov LISI*, Leningrad, No. 84. 51–62.

Klöppel, K. and Schubert, J. (1971). Die Berechnung der Traglast mittig und außermittig gedrückter, dünnwandiger Stützen . . . *Veröff. Inst. für Statik und Stahlbau TH Darmstadt*, Heft 13.

Knuth, D. E. (1975) Estimating the efficiency of backtrack programs. *Mathematics of Computation*, 29, 121–136.

Koch, K. F. (1973) Zur Anwendung von Verfahren der math. Optimierung beim Entwurf statisch unbestimmter Fachwerke. *Dissertation*. TU Braunschweig.

Konishi, Y. and Maeda, Y. (1976) Total cost optimum of I-section girders. *10th Congress of IABSE Tokyo. Preliminary Report*. 189–194.

Kreko, B. (1974) *Optimierung. Nichtlineare Modelle*. Budapest, Akadémiai Kiadó.

Krishnamoorthy, C. S. and Mosi, D. R. (1979) A survey on optimal design of civil engineering structural systems. *Engineering optimization*, 4, 73–88.

Kuester, J. L. and Mize, J. H. (1973) *Optimization techniques with Fortran*. New York, McGraw Hill.

Kunihiro, T., Saeki, S. and Inoue, K. (1976) Study on hybrid girders. *10th IABSE Congress Tokyo. Preliminary Report*, Zürich. 409–414.

Kunitskiy, L. P. (1978) Optimization and economy of hybrid beams (in Russian). *Soprotivlenie materialov i teoriya sooruzheniy. Vypusk 32*. Kiev, Budivel'nik. 17–25.

Kuzmanovic, B. O. and Willems, N. (1972) Optimum plastic design of steel frames. *J. Struct. Div. Proc. ASCE*, 98, 1697–1723.

Ladyzhenskiy, D. V. (1970) Calculation of the optimum height of trusses depending on the cross-section shape of elements and on the steel grade (in Russian). *Izv. Vuzov Stroit. i Arkhit.*, No. 2, 22–34.

Lanskoy, E. N. and Banketov, A. I. (1966) *Elementy rascheta i konstruktsii krivoshipnykh pressov*. Moskva, Masinostroenie.

LaPay, W. S. and Goble, G. G. (1971) Optimum design of trusses for ultimate loads. *J. Struct. Div. Proc. ASCE*, 97, 157–174.

Lawo, M. (1978) Optimization of stiffened plates. *Civ. Eng. Trans. Inst. Eng. Australia*, CE 20, 13–16.

Lawo, M., Murray, N. W. and Thierauf, G. (1978) Geschweißte gewichtsoptimierte I- und Kasten-Profile aus St 37. *Merkblatt 449*. Beratungsst. für Stahlverwendung, Düsseldorf.

Lee, S. L., Karasudhi, P., Zakeria, M. and Chan, K. S. (1971) Uniformly loaded orthotropic rectangular plate supported at the corners. *Civ. Eng. Trans. Inst. Eng. Australia*, 13, 101–106.

Lee, B. S. and Knapton, J. (1974) Optimum cost design of a steel-framed building. *Engineering Optimization*, 1, 139–153.

Lescouarc'h, Y. (1977) Capacité de résistance d'une section soumise à divers types de sollicitations. *Constr. Métall.*, 14, No. 2, 3–17.

Leśniak, Z. K. (1970) *Methoden der Optimierung von Konstruktionen unter Benutzung von Rechenautomaten.* Berlin–München–Düsseldorf, Ernst & Sohn.

Levey. G. E. and Fu, K. C. (1979) A method in discrete frame optimization and its outlook. *Computers & Structures*, 10, 363–368.

Levy, R. and Chai, K. (1979) Implementation of natural frequency analysis and optimality criterion design. *Computers & Structures*, 10, 277–282.

Lew, H. S., Natarajan, M. and Toprac, A. A. (1969) Static tests on hybrid plate girders. *Welding J.*, 48, s86–s96.

Lewis, A. D. M. (1968) Backtrack programming in welded girder design. *Proc. 5th Annual SHARE-ACM-IEEE Design Automation Workshop*, Washington. 28/1–28/9.

Liebman, J. S., Khachaturian, N. and Chanaratna, V. (1981) Discrete structural optimization. *J. Struct. Div. Proc. ASCE*, 107, 2177–2197.

Lin, J. H., Che, W. Y. and Yu, Y. S. (1982) Structural optimization on geometrical configuration and element sizing with statical and dynamical constraints. *Computers & Structures*, 15, 507–515.

Lipp, W. (1976) Ein Verfahren zur optimalen Dimensionierung allgemeiner Fachwerkkonstruktionen und ebener Rahmentragwerke. *Techn.-wiss. Mitt. Inst. konstruktiver Ingenieurbau Ruhr-Univ. Bochum*, No. 12.

Lipson, S. L. and Russell, A. D. (1971) Cost optimization of structural roof system. *J. Struct. Div. Proc. ASCE*, 97, 2057–2071.

Lipson, S. L. and Agrawal, K. M. (1974) Weight optimization of plane trusses. *J. Struct. Div. Proc. ASCE*, 100, 865–879.

Lipson, S. L. and Gwin, L. B. (1977) Discrete sizing of trusses for optimal geometry. *J. Struct. Div. Proc. ASCE*, 103, 1031–1046.

Lipson, S. L. and Haque, M. I. (1980) Optimal design of arches using the Complex method. *J. Struct. Div. Proc. ASCE*, 106, 2509–2525.

Little, G. H. (1979) The strength of square steel box columns – design curves and their theoretical basis. *Struct. Eng.*, 57A, No. 2, 49–61.

Maeda, Y. and Kawai, Y. (1972) Ultimate strength of longitudinally stiffened hybrid girders in bending. *Technol. Reports Osaka Univ.*, 22, No. 1050, 257–271.

Maeda, Y. and Konishi, Y. (1974–75) Optimum design including fabrication costs for I-section girders. *Technol. Reports Osaka Univ.*, 24, 317–324; 25, 217–223.

Maeda, Y., Ishiwata, M. and Kawai, Y. (1976) Structural behaviour of hybrid girders in bending. Application to actual bridges. *10th IABSE Congress Tokyo. Preliminary Report*, Zürich, 415–420.

Maier, G., Srinivasan, R. and Save, M. (1976) On limit design of frames using linear programming. *J. Struct. Mech.*, 4, 349–378.

Majid, K. I. (1974) *Optimum design of structures.* London, Newnes-Butterworths.

Majid, K. I. and Anderson, D. (1972) Optimum design of hyperstatic structures. *Int. J. Numer. Methods in Eng.*, 4, 561–578.

Majid, K. I., Stojanovski, P. and Saka, M. P. (1980) Minimum cost topological design of steel sway frames. *Struct. Eng.*, 58B, No. 1, 14–20.

Malkov, V. P. and Ugodchikov, A. G. (1981) *Optimizatsiya uprugikh sistem.* Moskva, Nauka.

Manual on stability of steel structures (1976) European Convention for Constructional Steelwork.

Maquoi, R. and Rondal, J. (1978) Mise en équation des nouvelles courbes européennes de flambement. *Constr. Métall.*, 15, No. 1, 17–30.

Markuš, S. and Valášková, O. (1972) On eigenvalue boundary problems of transversely vibrating sandwich beams. *J. Sound and Vib.*, 23, 423–432.

Markuš, S., Oravský, V. and Šimková, O. (1977) *Tlmené priečne kmitanie vrstvených nosníkov.* Bratislava, VEDA.

Massonnet, Ch. and Save, M. (1977) *Calcul plastique des constructions.* Vol. 1. 3. éd. Angleur-Liège, Nelissen, B.

Mathisen, Th. (1976) Valg av aveisemetode og materialer. *Maskin,* 9, No. 1, 7–8, 11.

McIntosh, S. C. (1974) Structural optimization via optimal control techniques: a review. *Structural optimization Symposium.* Ed. Schmit, L. A. ASME Winter Annual Meeting. AMD 7. 49–64.

Mead, D. J. and Markuš, S. (1969) The forced vibration of a three-layer damped, sandwich beam with arbitrary boundary conditions. *J. Sound and Vibr.*, 10, 163–175.

Miller, C. J. and Moll, Th. G. jr. (1979) Automatic design of tapered member gabled frames. *Computers & Structures,* 10, 847–854.

Mitra, G. P., Keshava Rao, M. N. and Gupta, A. K. (1978) Optimum design of lattice towers for power transmission and telecommunications. *J. Struct. Engng.*, 6, No. 1, 29–35.

Moe, J. and Lund, S. (1968) Cost and weight minimization of structures with special emphasis on longitudinal strength members of tankers. *Trans. Royal Inst. of Naval Arch,* 110, 43–70.

Moe, J. (1969) *Design of ship structures by means of nonlinear programming techniques.* Dept. of Ship Structures, Techn. Univ. of Norway, Trondheim.

Moe, J. (1974) Fundamentals of optimization. *Computers & Structures,* 4, 95–113.

Morgenstern, K. (1975) Materialsparende Auslegung von Pressengestellen. *Maschinenbautechnik,* 24, 486–490, 552–558.

Morris, A. J. (1972) Structural optimization by geometric programming. *Int. J. Solids and Struct.*, 8, 847–864.

Morris, L. J. and Randall, A. L. (1975) *Plastic design.* London, Constrado.

Moses, F. and Goble, G. G. (1970) Minimum cost structures by dynamic programming. *Engineering J.* (Amer. Inst. Steel Constr.) July.

Moshinskiy, S. I. (1976) Optimum design of thin-walled metal elements of statically indeterminate trusses (in Russian). *Gidromelioratsiya i gidrotekhnicheskoe stroitel'stvo. Vypusk 4.* L'vov, Vishcha shkola, 77–83.

Nakamura, T. and Nagase, T. (1976) Minimum weight plastic design of multi-story multispan plane frames subject to reaction constraints. *J. Struct. Mech.*, 4, 257–287.

Nelder, J. A. and Mead, R. (1965) A simplex method for function minimization. *Computer J.*, 7, 308–313.

Nelson, R. B. and Felton, L. P. (1972) Thin-walled beams in frame synthesis. *AIAA J.*, 10, 1565–1569.

Nethercot, D. A. (1976) Buckling of welded hybrid steel I-beams. *J. Struct. Div. Proc. ASCE,* 102, 461–474.

Niordson, F. and Olhoff, N. (1979) Variational methods in optimization of structures. *Reports Danish Center of Appl. Math. and Mech.* No. 161. 1–22.

Nuḍel'man, L. G. (1961) Investigation of the stiffness and strength of the hydraulic press P 474 (in Russian). *Kuznechno-shtamp. proizv.,* No. 7, 20–25.

Nudel'man, L. G. and Vereshchagin, Yu. F. (1963) Calculation method for welded frames of hydraulic presses (in Russian). *Nauka i proizvodstvo.* NTO Mashprom, Orenburg, Vypusk 1, 72–84.

Okerblom, N. O., Demyantsevich, V. P. and Baikova, I. P. (1963) *Proektirovanie tekhnologii izgotovleniya svarnykh konstruktsiy.* Leningrad, Sudpromgiz.

Olhoff, N. (1976) A survey of the optimal design of vibrating structural elements. *Danish Center Appl. Math. and Mech.* Report No. 102.

Ol'khov, N. (1981) *Optimal'noe proektirovanie konstruktsiy.* Moskva, Mir.

Ol'kov, Ja. I. and Antipin, A. A. (1978) Optimum material distribution in statically indeterminate pin-jointed rod systems (in Russian). *Izv. Vuzov. Stroit. i Arkhit.*, No. 6, 51–56.

Panc, V. (1959) Statika tenkostenných prutů a konstrukcí. Praha, ČSAV.

Pappas, M. (1980) An improved direct search numerical optimization procedure. *Computers & Structures*, 11, 539–557.

Pappas, M. (1981) Improved methods for large scale structural synthesis. *AIAA J.*, 19, 1227–1233.

Parimi, S. R. and Cohn, M. Z. (1978) Optimal solutions in probabilistic structural design. *J. méc. appl.*, 2, No. 1, 47–92.

Patel, J. M. and Patel, T. S. (1980) Minimum weight design of the stiffened cylindrical shell under pure bending. *Computers & Structures*, 11, 559–563.

Pavlov, V. A. and Shirinyants, V. A. (1979) Application of geometric programming to optimum design of structures (in Russian). *Uchenye Zapiski Tsentr. Aero-gidrodin. Inst.*, 10, No. 1, 99–104.

Pedersen, P. (1973) Optimal joint positions for space trusses. *J. Struct. Div. Proc. ASCE*, 99, 2459–2476.

Pedersen, P. (1980) The integrated approach of FEM-SLP for solving problems of optimal design. *Report Danish Center Appl. Math. and Mech.*, No. 182.

Pedersen, P. (1981) Integrating finite element and linear programming. *Report Danish Center Appl. Math. and Mech.*, No. S18.

Peschel, M. and Riedel, C. (1975) *Polyoptimierung.* Berlin, VEB Verlag Technik.

Peters, K. (1973) Kran mit Ovalträger. *Fördermittel J.* 5, No. 3, 66.

Pettersen, E. (1979) *Analysis and design of cellular structures.* Univ. of Trondheim, The Norwegian Inst. of Technology, Division of Marine Struct.

Pickett, R. M. jr., Rubinstein, M. F. and Nelson, R. B. (1973) Automated structural synthesis using a reduced number of design coordinates. *AIAA J.*, 11, 489–494.

Pierson, B. L. (1972) A survey of optimal structural design under dynamic constraints. *Int. J. Numer. Methods in Eng.*, 4, 491–499.

Plantema, F. J. (1966) *Sandwich construction.* New York, Wiley.

Plastic design in steel. 2nd ed. (1971) New York, Amer. Soc. Civ. Engrs.

Pochtman, Yu. M. (1975) Optimum design of structures by random search method (in Russian). *Problemy sluchaynogo poiska. Vypusk 4.* Red. Rastrigin, L. A. Riga, Zinatne, 184–193.

Polizzotto, C. and Mazzarella, C. (1979) Limit design of frames with axial force – bending moment interaction. *J. Struct. Mech.*, 7, 83–106.

Prager, W. (1956) Minimum weight design of a portal frame. *J. Engng. Mech. Div. Proc. ASCE*, 82, No. EM4.

Prasad, B. and Haftka, R. T. (1979a) A cubic extended interior penalty function for structural optimization. *Int. J. Numer. Methods in Eng.*, 14, 1107–1126.

Prasad, B. and Haftka, R. T. (1979b) Optimal structural design with plate finite elements. *J. Struct. Div. Proc. ASCE*, 105, 2367–2382.

Prasad, B. (1982) An improved variable penalty algorithm for automated structural design. *Computer Methods in Appl. Mech. and Eng.*, 30, 245–261.

Proc. Int. Coll. on Column Strength, Paris 1972. Reports of the Working Commissions Vol. 23. IABSE, Zürich, 1975.

Proc. 2nd Int. Coll. Stability of Steel Structures (1977) Preliminary Report. Liège.

Proc. Regional Colloquium Stability of Steel Structures (1977). (Eds Halász, O., Iványi, M.) Budapest.

Proc. Advanced Study Institute. (1981) Optimization of distributed parameter structures, Iowa, 1980. (Eds Haug, E. J., Cea, J.) Alphen ann de Rijn, Sijthoff-Nordhoff.

Proektirovanie i optimizatsiya konstruktsiy inzhenernykh sooruzheniy (1981) Red. Saltsevich, V. Ya. Riga, Politekhn. Inst.

Puchner, O. and Ruža, S. (1955) Měření na modelu svařovaného stojanu třecího lisu. *Zváranie,* **4,** 141–149.

Quinn, N. (1977) Factors which influence the cost of steel structures. *Proc. 2nd Conf. Steel Developments,* Austral. Inst. Steel. Constr. Melbourne, 92–99.

Ramakrishnan, C. V. and Francavilla, A. (1974–75) Structural shape optimization using penalty functions. *J. Struct. Mech.,* **3,** 403–422.

Rao, S. S. (1978) *Optimization. Theory and applications.* New Delhi, Wiley Eastern Ltd.

Rao, S. S. (1979) Structural optimization under shock and vibration environment. *Shock and Vibration Digest,* **11,** No. 2, 3–12.

Rao, S. S. and Reddy, E. S. (1980) Optimum design of stiffened cylindrical shells with natural frequency constraints. *Computers & Structures,* **12,** 211–219.

Rao, S. S. and Reddy, E. S. (1981) Optimum design of stiffened conical shells with natural frequency constraints. *Computers & Structures,* **14,** 103–110.

Rao, S. S. (1981) Reliability-based optimization under random vibration environment. *Computers & Structures,* **14,** 345–355.

Rastrigin, L. A. (1975) Application of the random search method to optimum design (in Russian). *Problemy sluchaynogo poiska.* No. 4, 7–17.

Reissner, E. and Weinitschke, H. J. (1963) Finite pure bending of circular cylindrical tubes. *Quart. Appl. Math.,* **20,** 305–319.

Reitman, M. I. and Shapiro, G. S. (1976) *Metody optimal'nogo proektirovaniya deformiruemykh tel.* Moskva, Nauka.

Reitman, M. I. and Shapiro, G. S. (1978) Optimum design of deformable solid bodies (in Russian). *Itogi Nauki i Tekhniki. VINITI. Ser. Mekh. Deform. Tverd. Tel,* **12,** 5–90.

Richter, G. (1974) Optimierung tragender Bauelemente, am Beispiel geschweisster T, Doppel-T und Kastenprofile. *Dokt. Dissertation.* Techn. Hochschule Karl-Marx-Stadt.

Rizzi, P. (1976) The optimization of structures with complex constraints via a general optimality criteria method. *Ph. D. Thesis,* Stanford Univ. California.

Rockey, K. C. and Škaloud, M. (1972) The ultimate load behaviour of plate girders loaded in shear. *Struct. Eng.,* **50,** No. 1, 29–47.

Rockey, K. C., Evans, H. R. and Porter, D. M. (1978) A design method for predicting the collapse behaviour of plate girders. *Proc. Inst. Civ. Engrs Part 2.,* **65,** Mar. 85–112.

Rockey, K. C. and Valtinat, G. (1978) Zur Traglastberechnung vollwandiger Biegeträger mit Vertikalsteifen. *Integration von Maschinen- und Stahlbau.* Mainz, Krausskopf-Verlag, 95–112.

Rondal, J. and Maquoi, R. (1977) Optimization of unstiffened hybrid I-beams with stability constraints. *Proc. Regional Coll. Stability of Steel Struct.* Budapest, 373–382.

Rondal, J. and Maquoi, R. (1979) Formulations d'Ayrton–Perry pour le flambement des barres métalliques. *Constr. Métall.,* **16,** No. 4, 41–53.

Rosen, A. and Schmit, L. A. (1981) Optimization of truss structures. *AIAA J.,* **19,** 626––634.

Rosenkranz, B. (1968) Contributo alla ottimizzazione del peso di travi a traliccio. *Constr. Metall.,* **20,** 329–338.

Saka, M. P. (1980a) Shape optimization of trusses. *J. Struct. Div. Proc. ASCE,* **106,** 1155––1174.

Saka, M. P. (1980b) Optimum design of rigidly jointed frames. *Computers & Structures,* **11,** 411–419.

Schade, H. A. (1941) Design curves for cross-stiffened plating under uniform bending load. *Trans. Soc. Nav. Arch. and Marine Engrs,* **49,** 154–182.

Schilling, Ch. G. (1974) Optimum properties for I-shaped beams. *J. Struct. Div. Proc. ASCE,* **100,** 2385–2401.

Schmit, L. A. jr. (1960) Structural design by systematic synthesis. *Proc. 2nd Conf. Electronic Computation,* ASCE, New York, 105–132.

Schmit, L. A. and Miura, H. (1976) A new structural analysis-synthesis capability – ACCESS 1. *AIAA J.,* **14,** 661–671.

Schmit, L. A. and Miura, H. (1978) An advanced structural analysis-synthesis capability ACCESS 2. *Int. J. Numer. Methods in Eng.,* **12,** 353–377.

Schmit, L. A. and Ramanathan, R. K. (1978) Multilevel approach to minimum weight design including buckling constraints. *AIAA J.,* **16,** 97–104.

Schmit, L. A. and Fleury, C. (1980) Structural synthesis by combining approximation concepts and dual methods. *AIAA J.,* **18,** 1252–1260.

Schmit, L. A. (1981) Structural synthesis – its genesis and development. *AIAA J.,* **19,** 1249–1263.

Schweer, W. and Mewes, W. (1969) Beitrag zur Frage geschweißter oder gegossener Pressenständer. *Blech,* **16,** 116–121.

Seaburg, P. A. and Salmon, C. G. (1971) Minimum weight design of light gage steel members. *J. Struct. Div. Proc. ASCE,* **97,** 203–222.

Sergeev, N. D. and Bogatyrev, A. I. (1971) *Problemy optimal'nogo proektirovaniya konstruktsiy.* Leningrad, Stroyizdat.

Shamie, J. and Schmit, L. A. jr. (1975) Frame optimization including frequency constraints. *J. Struct. Div. Proc. ASCE,* **101,** 283–293.

Shanley, F. R. (1960) *Weight-strength analysis of aircraft structures.* 2nd ed. New York, Dover.

Shanmugam, N. E. (1978) The design of multi-cell structures. *Ph. D. Thesis.* Univ. College, Cardiff.

Shaw, R. (1974) Minimum weight design of structures with frequency constraints. *Dissertation.* The Pennsylvania State Univ.

Shimada, W., Hoshinouchi, S., Hiramoto, S., Hijikata, A., Yoshioka, S. and Inoue, A. (1978) Improvement of fatigue strength in fillet welded joint by CO_2 soft plasma arc dressing of weld toe. *IIW doc.* XIII-881-78.

Shive, A. R. (1972) Minimum weight design of steel girders. *Ph. D. Dissertation.* Rice Univ. Houston, Texas.

Siddall, J. N. (1972) *Analytical decision-making in engineering design.* Prentice-Hall, Englewood Cliffs, N. J.

Sikalo, P. I. (1977) Practical methods for the minimum weight design of welded hybrid I-beams (in Russian). *Metallicheskie konstruktsii i ispytaniya sooruzheniy.* Leningrad, LISI, No. 1 (134), 79–87.

Sikalo, P. I. (1978) Minimum mass design of welded plate girders using stress and stiffness constraints (in Russian). *Metallicheskie konstruktsii i ispytaniya sooruzheniy.* Leningrad, LISI, 14–21.

Škaloud, M. (1962) Interaktion der Ausbeulung von Wänden und der gesamten Formänderung gedrückter und gebogener Stäbe. *Acta Techn. ČSAV,* No. 1. 52–.

Škaloud, M. and Zörnerová, M. (1970) Experimental investigation into the interaction of the buckling of compressed thin-walled columns with the buckling of their plate elements. *Acta Techn. ČSAV,* 389–423.

Škaloud, M. and Náprstek, J. (1977) Limit state of compressed thin-walled steel columns with regard to the interaction between column and plate buckling. *Proc. 2nd Int. Coll. Stability of Steel Struct. Liège. Preliminary Report,* 405–414.

Snowdon, J. C. (1968) *Vibration and shock in damped mechanical systems.* New York, Wiley.

Spendley, W., Hext, G. R. and Himsworth, F. R. (1962) Sequential application of simplex designs in optimization and evolutionary operation. *Technometrics,* **4,** 441–461.

Spunt, L. (1971) *Optimum structural design.* Prentice-Hall, Englewood Cliffs, N. J.

Stability of metal structures – a world view. (1981–82) *Engineering J.* (AISC), 18, 90–125, 154–196; 19, 27–62, 101–138.

Stainsby, R. (1980) Automation in the fabrication and design of welded beams, girders and columns. *Struct. Eng.* 58A, No. 5, 149–153.

Stamm, K. and Witte, H. (1974) Sandwichkonstruktionen. Berlin, Springer.

Svanberg, K. (1981) Optimization of geometry in truss design. *Computer Methods in Appl. Mech. and Eng.*, 28, 63–80.

Svensson, S. E. and Croll, J. G. A. (1975) Interaction between local and overall buckling. *Int. J. Mech. Sci.*, 17, 307–321.

Szabó, J. and Roller, B. (1978) *Anwendung der Matrizenrechnung auf Stabwerke.* Budapest, Akadémiai Kiadó.

Szilard, R. (1974) *Theory and analysis of plates.* Prentice-Hall, Englewood Cliffs, N. J.

Tabak, E. I. and Wright, P. M. (1981) Optimality criteria method for building frames. *J. Struct. Div. Proc. ASCE*, 107, 1327–1342.

Templeman, A. B. (1970) Structural design for minimum cost using the method of geometrical programming. *Proc. Inst. Civ. Eng.*, 46, 459–472.

Thomas, H. R. jr. and Brown, D. M. (1977) Optimum least-cost design of a truss roof system. *Computers & Structures*, 7, 13–22.

Tímár, I. (1981) Optimierung von Sandwichplatten mit der nichtlinearen SUMT-Methode. *Konstruktion.* 33, 403–407.

Timerbaev, N. S. (1973) Experimental investigation of the plastic buckling of pipes subjected to bending (in Russian). *Trudy VNII po sboru, podgotovke i transporta nefti i nefteproduktov, Vyp. 11,* 165–173.

Timoshenko, S. and Woinowsky-Krieger, S. (1959) *Theory of plates and shells.* 2nd ed. New York, McGraw-Hill.

Toakley, A. R. (1968) Optimum design using available sections. *J. Struct. Div. Proc. ASCE*, 94, 1219–1241.

Toakley, A. R. and Williams, D. G. (1977) The optimum design of stiffened panels subject to compression loading. *Engineering Optimization*, 2, 239–250.

Toprac, A. A. and Natarajan, M. (1971) Fatigue strength of hybrid plate girders. *J. Struct. Div. Proc. ASCE*, 97, 1203–1226.

Turner, H. K. and Plaut, R. H. (1980) Optimal design for stability under multiple loads. *J. Eng. Mech. Div. Proc. ASCE;* 106, 1365–1382.

Uhlmann, W. (1979) Ausgewählte Rahmenformeln für das Traglastverfahren. Düsseldorf, Ernst.

Ungar, E. E. (1962) Loss factors of visco-elastically damped beam structures. *J. Acoust. Soc. Amer.*, 38, 1082–1089.

Usami, T. and Fukumoto, Y. (1982) Local and overall buckling strength tests of high strength steel box columns. *J. Struct. Div. Proc. ASCE*, 108, 525–542.

Vachajitpan, P. and Rockey, K. C. (1977) Optimum design of stiffened plate girders. *Large engineering systems. Proc. Int. Symposium,* Winnipeg, 1976. Oxford, e. a. 248–259.

Vandepitte, D. and Rathé, J. (1980) Buckling of circular cylindrical shells under axial load in the elastic-plastic region. *Stahlbau,* 49, 369–373.

Vanderplaats, G. N. and Moses, F. (1972) Automated design of trusses for optimum geometry. *J. Struct. Div. Proc. ASCE*, 98, 671–690.

Venkayya, V. B. (1971) Design of optimum structures. *Computers & Structures,* 1, 265–309.

Venkayya, V. B., Khot, N. S. and Berke, L. (1973) Application of optimality criteria approaches to automated design of large practical structures. *Proc. 2nd Symposium on Structural Optimization AGARD–CP–123,* Milan, 3/1–3/19.

Venkayya, V. B. (1978) Structural optimization: a review and some recommendations. *Int. J. Numer. Methods in Eng.*, **13**, 203–228.

Verner, D. (1976) Vibrations excited by people (in Russian). *Stroitel'naya Mekhanika i Raschet Sooruzheniy*, No. 1, 58–59.

Vershinskiy, A. V. and Ryadnova, L. V. (1979) Methods for the weight reduction of over-head travelling cranes (in Russian). *Teoriya, raschet i issledovanie PTM. Trudy MVTU No. 315.* Moskva, 26–62.

Vilnay, O. and Rockey, K. C. (1981) A generalised effective width method for plates loaded in compression. *J. Constructional Steel Research*, **1**, No. 3, 3–12.

Vilnay, O. (1983) A full orthotropic-plate method for double-bottom structures. *J. Constructional Steel Research*, **3**, No. 1, 19–27.

Volkov, V. V. (1978) Determination of the manpower requirements in the production of the structural parts of industrial buildings in the design process (in Russian). *Issledovanie tekhnologii izgotovleniya metallicheskikh konstruktsiy. Trudy CNII Proektstal'konstruktsiya, Vypusk 23.* 34–45.

Vol'mir, A. S. (1967) *Ustoychivost' deformiruemykh sistem.* Moskva, Nauka.

Voprosy optimal'nogo proektirovaniya plastin i obolochek (1981) Saratov, Saratovsk. Univ.

Voronin, V. G. (1966) Frame structures of hydraulic presses (in Russian). *Vestnik Mashinostroeniya*, **46**, No. 5, 61–71.

Walker, R. J. (1960) An enumerative technique for a class of combinatorical problems. *Proc. of Symposia in Appl. Math. Amer. Math. Soc.* Providence, R. I., **10**, 91–94.

Wardenier, J. (1982) *Hollow section joints.* Delft, Delft Univ. Press.

White, J. D. (1977a) *Longitudinal shrinkage of a single pass weld.* Univ. of Cambridge, Dept. of Engng. CUED/C-Struct./TR. 57.

White, J. D. (1977b) *Longitudinal stresses in a member containing non-interacting welds.* Univ. of Cambridge, Dept. of Engng. CUED/C-Struct./TR. 58.

White, J. D. (1977c) *Longitudinal shrinkage of multi-pass welds.* Univ. of Cambridge, Dept. of Engng, CUED/C-Struct./TR. 59.

White, J. D. (1977d) *Longitudinal stresses in welded T-sections.* Univ. of Cambridge, Dept. of Eng. CUED/C-Struct./TR. 60.

Wills, J. (1973) A mathematical optimization procedure and its application to the design of bridge structures. *Dept. of the Environment, Transport and Road Research Lab. Report LR 555.* Crowthorne.

Wright, P. M. and Hakim, H. F. (1978) Automated design of rigid steel frames including member selection. *Canadian J. Civ. Eng.*, **5**, 114–125.

Yamasaki, T., Hara, M. and Kawai, Y. (1976) Fatigue strength of longitudinal fillet welded joints in hybrid girders. *10th IABSE Congress Tokyo. Preliminary Report, Zürich*, 415–426.

Yin, T. P., Kelly, T. J. and Barry, J. E. (1967) A quantitative evaluation of constrained-layer damping. *J. Eng. for Industry, Trans. ASME*, **89**, 773–784.

Yoshida, H. and Maegawa, K. (1979) The optimum cross-section of channel columns. *Int. J. Mech. Sci.*, **21**, 149–160.

Yoshimura, M. (1980) Optimum design of machine structures . . . *Bull. Japan Soc. Precis. Eng.*, **14**, 236–242.

Zherbin, M. M. (1974) *Vysokoprochnye stroitel'nye stali.* Kiev, Budivel'nik.

Design standards and rules

BS 5400:Part 3: 1982. Code of practice for design of steel bridges. British Standard Inst.

ČSN 731401 Czechoslovak design standard for steel structures.

ČSN 270103 Czechoslovak design standard for steel structures of cranes, 1970.

DIN 4132 (1981). Kranbahnen. Stahltragwerke.

ISO 657/XIV–1977. Hot-rolled steel sections. Part XIV: Hot-finished structural hollow sections.

ISO/DIS 4019.2 Draft Int. Standard. Cold-finished steel structural hollow sections, 1979.

MI–04–188–80. Hungarian plastic design concepts for steel structures in building industry. 1980.

MSZ 15024/1–75. Hungarian design standard for steel structures. 1975.

SIA 161. Stahlbauten. Schweizerischer Ingenieur- und Architekten-Verein, Zürich, 1979.

SNIP II–23–81. Soviet design standard for steel structures. Stal'nye konstruktsii. Gosstroy, Moskva. 1982.

Specification for the design, fabrication and erection of structural steel for buildings. Amer. Inst. Steel Constr. 1969.

Index